An Introduction to U.S. Patent Searching

An Introduction to U.S. Patent Searching
The Process

Susan B. Ardis

1991
LIBRARIES UNLIMITED, INC.
Englewood, Colorado

LIBRARIES UNLIMITED, INC.
P.O. Box 3988
Englewood, CO 80155-3988

Library of Congress Cataloging-in-Publication Data

Ardis, Susan.
An introduction to U.S. patent searching : the process / Susan B. Ardis.
xv, 221 p. 22x28 cm.
Includes bibliographical references and index.
ISBN 0-87287-856-2
1. Patent searching. I. Title. II. Title: Introduction to United States patent searching.
T210.A73 1991
608.773--dc20 91-14507
CIP

To Tom, Jessica, and Andrew,
who listened and supported this project.

The U.S. patent system was established by the wisdom of our forefathers with the intent of stimulating invention by providing suitable incentive. The purpose of stimulating invention was primarily to benefit the public.

—Dr. Augustus Kinzel
Proceedings of ad hoc committee on the Role of Patents,
National Research Council, 1962.

Contents

Appendixes — *Continued*

Introduction

The greatest invention of the nineteenth century was the invention of the method of invention.

—Alfred North Whitehead

PATENTS: FUEL FOR THOUGHT

Patents are by definition an open source of information, to be freely read and examined by the public. The U.S. patent system has long been recognized as a vast and growing resource of technical knowledge. Found in this collection are solutions to all types of real, and imagined, technical problems. These solutions range from practical to esoteric to fanciful. What sets patent information apart from other information resources is its organization. It is one of the most thoroughly indexed and documented forms of literature, allowing specific information to be traced, reviewed, and retrieved with considerable accuracy.

Patents cover a wide range of topics, including mechanical and electrical inventions, pharmaceuticals and chemicals, asexually reproduced plants, and ornamental designs. Historically, the primary use of patents has been to determine if something has already been patented. However, patents can also be used by historians, economists, sociologists, and others for such varied purposes as examining the development of specific areas of technical growth, researching the commercial interests of selected companies or the expertise and creativity of specific individuals, or for market research. In other words, the patent file offers a unique historical collection of information on a wide range of topics. While patents can give clues to historical developments, their primary use is as a file of current technological solutions and advances. The file can yield information to stimulate new ideas or to avoid wasting time and money reinventing the wheel.

The nine most common reasons, or occasions, for conducting a patent search are:

1. Preliminary searches for information about a new field of interest or possible research.
2. Information searching by researchers working on specific projects.
3. Searches done prior to filing a patent application.
4. Searches done prior to taking a license under another inventor's patent.
5. Opposition searches as part of infringement cases, or other patent litigation.
6. Information searches leading to development of new or related products and processes.
7. Searches for market information: the type and number of patents reflect current activity in certain area of interest by company name.
8. Competitor tracking: these searches can reveal patent activities of competitors.
9. Technology tracking: these allow companies to spot new competitors and to identify technological trends.

Given the depth and size of this reservoir of knowledge the light use and low value given it by the public are surprising. Despite numerous recent articles encouraging the use of patents, their actual use remains low and their potential value is still unrealized. This low use is often attributed to ignorance, to fear of a system which seems to be highly complicated, and to difficulty in reading and understanding the legal language of the documents. Many people probably do not use patents because the value of the documents has not been clearly explained to them and because they have had no training in the use of patents.

While some groups and individuals are encouraging use of this resource, others disparage the importance of both patents and the patent literature by saying that if an invention is important it will appear in the journal literature. But according to John Terapane (1978), over 70 percent of inventions described in patents never appear in the journal literature, and those that do often appear years later. A good example is the invention of the punched card. It was patented twenty-five years before it was described in a journal article.

Many conflicting statements have been made over the years by scientists and company or government officials about the importance of filing for patents. Some have even suggested that secrecy gives more effective protection. Others have suggested that the Patent and Trademark Office is too restrictive in allowing claims and too slow in granting patents (Fox 1981; Samuel 1983; Fox 1984). In light of these suggestions, Stuart Kaback has made an extraordinary statement: "If I were told that there could be only one type of publication, either patents or scientific journals, I'd take patents without hesitation" (Kaback 1978).

No one can deny that patented inventions have played an important role in our everyday lives. In fact, it is hard to imagine what our lives would be like without U.S. patent 1,647, granted to Samuel F. B. Morse in 1840 for "Improving the Mode of Communicating Information by Signals by the Application of Electro-Magnetism" (telegraph); U.S. patent 174,465, granted to Alexander Graham Bell in 1876 for an "Improvement in Telegraphy" (telephone); U.S. patent 504,038, granted to Whitcomb L. Judson in 1893 for a "Clasp Locker or Unlocker for Shoes" (zipper); or U.S. patent 2,221,776, granted to Chester F. Carlson in 1940 for "Electron Photography" (Xerography-photocopying).

One of the purposes of this book is to stimulate interest in and to encourage use of patents as an information source whether or not the researcher's primary goal is to file a patent application. The major goal of this book is to facilitate use of patents by explaining how patents are organized, how this organization affects retrieval, how information is actually retrieved, and what elements constitute a patent document. The skills necessary to retrieve patent information include use of the *Index to the U.S. Patent Classification*, the *Manual of Classification*, the *Classification Definitions*, the *Assignee Listing*, and the various online systems.* To stimulate use this book attempts to provide an overview of the patent literature as well as a step-by-step guide to performing patent searches.

Librarians frequently receive requests for assistance from bewildered inventors seeking either patents or help/direction in starting a patent search. All too often librarians share this bewilderment and are overwhelmed by the seeming complexity of the material, while potential patentees are inhibited by the expense involved in hiring a professional searcher or patent attorney. However, both inventors and librarians can benefit from doing preliminary research. Determining the patentability of an idea or the desirability of undertaking a full-blown search should start with checking at the local public or college library to see if it is either a U.S. government publications depository or a patent depository library. All patent depository libraries and many public libraries receive the same basic patent search tools as part of their depository status.** These tools include the *Official Gazette, Index to the U.S. Patent Classification, Classification Definitions*, and the *Manual of Classification*.

The only way to become familiar with the process and its complexity is to actually experience it. Therefore, descriptions and explanations of the search process are included with sample searches. These searches have been included to demonstrate usage of the tools, search techniques, and search procedures. Each was selected because it demonstrated either specific problems describing modern technology or the

---

*A complete list of patent search tools, with annotations, is found in chapter 14.

**A complete listing of all U.S. patent depository libraries is found in appendix 13.

type of search commonly needed. Each sample search is presented step-by-step. Because patent terminology is complex and arcane, readers should consult the glossary for definitions as they proceed through the book.

Following an actual search step-by-step through the process is very valuable and enlightening; however, to be confident, searchers should perform several searches on their own. Consequently a number of search problems have been included for practice. Answers are found in chapter 13. By following the examples and working the problems, researchers will learn to find and retrieve information contained in patents.

REFERENCES

Fox, B. 1981. "Is the Day of the Patent Over?" *New Scientist* (Sept. 10): 653-55.

Fox, J. R. 1984. "Patents are Encroaching on Research Freedom." *Science* 224 (June 8): 1080-92.

Kaback, S. 1978. "Retrieving Patent Information Online." *Online* 2, 1: 16-25.

Samuel, R. I. 1983. "The Problems with Patents." *Machine Design* 55 (Sept. 8): 182-83.

Terapane, J. 1978. "Using U.S. Patents." *Chemtech* (May): 272-76.

1 Inventing the Patent

The American Patent System has promoted countless applications of the arts and sciences to the needs and well-being of our people.

—Franklin D. Roosevelt

ORIGIN AND EARLY HISTORY

Patents have been granted in one form or another for thousands of years. However, what constitutes a patent has changed dramatically over the centuries. Originally a *patent* was defined as any grant from a sovereign of a special license or privilege issued in the form of open letters addressed to the public at large. The earliest recorded examples of a special license or privilege were those granted to cooks in ancient Greece who developed special recipes. The word *patent* is derived from the Latin name for these documents, *literae patentes*, or letters patent. This phrase was commonly abbreviated to "patent."

Early patents were not necessarily connected to technology; they included charters, commissions, titles of nobility, grants of monopoly, and authority to conduct explorations. They were issued and abrogated according to royal whim and were often used as rewards or to pay off debts. As a result, court favorites and creditors were often given monopolies over such existing products and technologies as the making of salt or paper. Unlike patents today, these early ones did not imply the creation or introduction of something unique, because they were designed to reward favorites rather than to stimulate commerce or industry. In fact, the intent was often to retard commerce. Nevertheless, these grants are the direct predecessors of today's patent.

THE MIDDLE AGES AND RENAISSANCE

The nature of early letters patent began to change in the fourteenth century, when English monarchs developed a form of state-granted monopoly or "letters of protection" to induce foreign craftsmen to emigrate. The first craftsmen to accept this offer were two weavers from Brabant who immigrated to England in 1331. The first inventor known to participate was John of Shiedame, who brought a new method for producing salt to England in 1440 (Martin 1905). Another example of an early patent for an invention was that granted by the Doge of Venice in 1594 to Galileo for a "machine for raising water and irrigating land" (Goodeve 1884).

The policies relating to patents or state monopolies developed in a haphazard manner until the early years of the reign of Queen Elizabeth I, when the bankruptcy of the royal treasury and the government's need for cash forced her to encourage national industry. Elizabeth is also credited with initiating the beginnings of a modern patent system. These policies were probably related to the numerous economic changes taking place in England and on the Continent in the 1500s. The guild system was declining, industry was becoming national rather than local, and a capitalist middle class was beginning to form. Previously inventors and manufacturers had protected their processes and inventions by keeping them secret. The great risks associated with new forms of commerce and the decline of the guild system encouraged the idea of granting economic privileges. Newly formed companies, under the direction of the crown but at their own expense, undertook the formation of needed industries. One of the rewards for their risk was to receive practical monopolies or patents.

An early statement of the economic benefit of patents was recorded in a patent petition by Acontion in 1559. In the petition, he stated that he had invented certain furnaces and "wheeled-machines" that others would copy to his loss unless he were protected. His petition also contained the first statement of why inventors should be rewarded with a monopoly: "Nothing is more honest than that those who by searching have found out things useful to the public should have some fruits of their rights and labors, as meanwhile they abandon all other modes of gain, are at much expense in experiments, and often sustain much loss" (Prager 1944).

Early in the seventeenth century, two key principles of the modern patent system were added to the patent structure. First, the Lord Chief Justice of England, Edward Coke, espoused the idea that patents for inventions deprived society of nothing, and thus could not possibly be harmful monopolies. Since an invention is by definition something new, argued Coke, society has everything to gain and nothing to lose by granting exclusive privileges to the inventor. Abuses of other forms of monopolies brought the second principle into effect: the government stipulated that monopoly rights could be awarded only to those who created something unique (Coke 1797-1823).

MODERN-DAY PATENT SYSTEM

These two principles are the foundation of the modern patent system. In the United States the phrase "patent system" covers almost everything concerning the rights of inventors, including legal custom and practices, acts of Congress, opinions of the Supreme Court, and activities of the Patent Office. The system attempts to ensure that the public interest is served by encouraging commerce, rewarding individual invention, and providing for public disclosure of the main elements of the invention.

In building a patent system to encourage new developments, the United States borrowed freely from England. Thus, as in England, service to the public, rather than to the individual, became the principal justification for patents.

After the Revolutionary War, exclusive privileges continued to be granted to inventors by the individual states. These state grants were part of the transition period connecting the earlier colonial monopolies to the modern patent system. These early state patents were granted by the individual legislatures, operating on the principle that anyone who had invented something unique was entitled to a patent as a right rather than as a special grant. With a few exceptions, the states continued to issue patents until the adoption of the Constitution and the enactment of the Patent Act of 1790. The Constitution grants to Congress "the power ... to promote the progress of science and the useful arts, by securing for limited times to authors and inventors the exclusive right to their respective writings and discoveries" (*Constitution* 1789).

During the first session of the first Congress, eighteen individual petitions were received, most of which were for patents. It became evident that neither Congress nor the inventor should be burdened with the lengthy and uncertain action of an individual law for each case, and that the whole matter was better suited to an administrative machinery. A bill reported on June 23, 1789, provided for both patents and copyright. George Washington underlined the importance of this measure when he recommended to the second session of the first Congress the value of encouraging invention: "I cannot forbear intimating to you the expediency of giving effectual encouragement, as well to the introduction of new and useful inventions from abroad, as to the exertions of skill and genius in producing them at home" (Patent Laws 1932).

The patent bill, which did not include copyright, was passed by both houses of Congress and was signed by the president on April 10, 1790, little more than one year after the organization of the new government. Subject matter for a patent was defined as "any useful art, manufacture, engine, machine, or device, or any improvement therein not before known or used" (Patent Laws 1932). The inventor was to present a petition to a board made up of the secretary of state, the secretary of war, and the attorney general, who were empowered to grant the patent in the name of the United States, "if they shall deem the invention or discovery sufficiently useful and important." The inventor was required to file a written specification, a drawing, and a model if possible. These were to be "so particular and so exact that the invention could be distinguished from other things before known and used, and that persons skilled in the art could make, construct or use the invention" (Patent Laws 1932). It is interesting that no oath was required.

Fifty-seven patents were granted under the first patent act, which was superseded by a second act in 1793. Not much information about the subject matter of these first patents is available because all records were destroyed in the Patent Office fire of 1836.

Since the first patent statutes were enacted in the 1790s, U.S. patent law has undergone only a few serious or basic changes. The Patent Act of 1793 enlarged on the previous statute by instituting an oath, establishing procedures for handling interfering applications, omitting the requirement that the board "deem the invention or discovery sufficiently useful and important," and clarifying the nature of the rights conferred on a patent holder.

These rights provided that the inventor of an improvement in something previously patented and still in force could not make, use, or sell the original discovery. Neither could the first patentee use the improvement. This provision was inherent in the original act, since a patent does not confer a positive right to make, use, or sell anything, but does prevent others from doing these things. The act stated that "changes in form or proportions shall not be deemed a discovery."

Difficulties with both laws led to passage in 1836 of a new patent act. The new act created a systematic examination method of granting patents, established a separate Patent Office, and provided the organization necessary to predetermine the validity of any invention before a patent was issued. For the first time it was possible to ascertain and protect the property rights of inventors in an intelligent and scientific way. This act provided the means for collecting the art (*inventions*), personnel trained to examine the art (*examiners*), and the procedures for scientifically fulfilling the law. The current numbering system was also initiated.* By 1836 10,000 applications had matured into patents organized into twenty-two classes. The 1836 law provided for a dignified procedure and granted prima facie evidence of the patent's validity.

The next major event occurred in 1870, and it resulted in better organization of the information. The principal purpose of this new act was to revise and consolidate the earlier statutes and to authorize the Commissioner to print copies of current patents as issued, and to "publish such laws, decisions, and rules as were necessary for the information of the public." On January 11, 1871, a joint resolution in Congress ordered the discontinuation of the old Patent Office Reports and directed the printing of copies of all patents. Previously searchers had had to hand copy the models and drawings stored in the Patent Office. In 1872 the first weekly issue of the *Official Gazette of the United States Patent and Trademark Office* (*OG* 1872-) appeared. Since then the *Official Gazette* has systematically recorded excerpts from patents and the more important Patent Office decisions. These excerpts have generally come from the patent claims, but recently only the largest claim, together with the most appropriate figure (drawing), have been printed. One interesting sidelight of the 1870 act is that it made the first statutory provision for the registration of trademarks.

---

*The first patent granted under the new numbering system was to John Ruggles, for a "Locomotive Steam Engine for Rail and Other Roads." The first patent actually granted in the United States was to Samuel Hopkins on July 31, 1790, for the "Making of Potash and Pearl-ash by new apparatus and Process."

Between 1776 and 1872 approximately 131,000 patents were issued. In 1872 these patents were reorganized into 145 classes to facilitate examination and provide public access to the technical information found in patents. This system continued to grow spasmodically, since there were no procedures for creating new classes or for the placement of patents in existing classes. This problem was solved to some degree in 1900 with the creation of specific written procedures for establishing new classes. These new procedures also described the methodology used to create the classes and the theoretical basis used to organize the information into reasonably searchable units. The last definitive effort to document the classification system occurred in 1966, with the publication of the *Development and Use of Patent Classification Systems* (Department of Commerce 1966).

Numerous revisions and consolidations have been made to the patent statutes since 1900, but probably the most important occurred on May 23, 1930, when President Herbert Hoover signed the act that added design and plant patents to what can be patented. As a result, soon afterward the basics of what can be patented and the organization of the patent file system were in place.

The last most important change occurred in the 1980s, when the U.S. Supreme Court decided that biotechnology and its products were patentable. As a result, the first patent on a biologically altered organism was issued April 12, 1988, to Philip Leder for "Transgenic Non-Human Mammals" (U.S. patent 4,736,866).

CONCLUSION

Patents have had a long and varied history, and they continue to play an important role in U.S. industry. As research and development grow in importance the patent system should continue to protect and encourage invention. Knowing the history of patents will probably not make them any easier to use, but it may provide some understanding of why the system is the way it is.

The following chapters provide basic background and information on what can be patented, the parts of the patent document, use of the patent tools, and how to begin a patent search.

REFERENCES

Buckles, R. A. 1957. *Ideas, Inventions, and Patents*. New York: John Wiley.

Coke, Sir Edward. 1797-1823. *Institutes of Law*. 18th ed. London: (n.p.).

Constitution of the United States of America. 1789. Article 1, Section 8, Clause 8.

Department of Commerce. 1966. *Development and Use of Patent Classification Systems*. Washington, D.C.: Government Printing Office.

Goodeve, T. M. 1884. *Cases Relating to Letters Patent for Inventors*. London: (n.p.).

Gordon, J. W. 1897. *Monopolies by Patents*. London: (n.p.).

Macaulay, Thomas Bobington. 1849. *The History of England*. 5th ed. London: (n.p.).

Martin, W. 1905. *English Patent System*. London: (n.p.).

Official Gazette of the United States Patent and Trademark Office (OG). 1872- . Washington, D.C.: U.S. Patent and Trademark Office.

Patent Laws, General Revision of. 1932. "Message to First Congress." Washington, D.C.: (n.p.).

Prager, F. D. 1944. *Journal of the Patent Office Society*. 26: 711-28.

ADDITIONAL READINGS

"Origin and Early History of Patents." 1936. *Journal of the Patent Office Society* 18, 7: 19-34.

"Outline of the History of the United States Patent Office." 1936. *Journal of the Patent Office Society* 18, 7.

Skolnik, H. 1979. "Historical Aspects of Patent Systems." *IEEE Transactions on Professional Communications* PC-22, 2: 59-63.

2 Questions, Answers, and Examples

I have seen with real alarm, several recent attempts in quarters carrying some authority to impugn the principle of patents altogether—attempts which, if practically successful, would enthrone free stealing under the prostituted name of free trade.

—John Stewart Mill

Provided in this chapter are basic information on the nature of patents; the statutory classes; the types of patents; areas of confusion with patents, such as copyrights; and definitions of some of the concepts and terms used by the Patent Office (see also the glossary). The U.S. patent system is complex, and background information provided in this chapter can help the researcher put the task in perspective.

The Patent Office, for a time, published a brief, nontechnical pamphlet called "Q and A About Patents" to answer the most frequently asked patent questions. The question-and-answer format dealt effectively with the vast amount of basic information needed, and was also a good way to define what is patentable. This format is used in this chapter and enhanced by recent examples, where relevant, at the end of each answer.

What Is a Patent?

A patent is a right of ownership granted by the government to a person, partnership, or corporation. A patent gives the owner the right to exclude others from making, using, or selling the "claimed" invention throughout the United States for seventeen years, provided the relevant fees are paid. It does *not* grant the inventor the unrestricted right to make or sell the invention, if doing so would infringe on the rights of others.

Example: Let us assume that you have invented a mechanism that allows both the umbrella's ribs and the handle to collapse into a compact package, and let us also assume that the original patent for the umbrella is still in effect. Would you be allowed to manufacture your invention? No, not if your invention infringed (used) the prior invention. You could, however, manufacture your invention if you had an agreement (license) to do so with the owner of the umbrella patent. On the other hand, when the umbrella patent expired, you and others would be able to use it without repercussions. However, others would not be able to add your collapsing mechanism until your patent expired.

Who Can Apply for a Patent?

Anyone can apply for a patent regardless of age, nationality, mental ability, or any other characteristic, as long as that person is the true inventor.

Example: Russian-born Vladimir Zworykin was granted U.S. patent 2,139,296 in 1938 for the "Cathode Ray Tube."

What Can Be Patented?

Basically, there are three different types of patents: utility, design, and plant. Each type has its own numbering sequence. Examples of each type are found in appendixes 1, 2, and 3, respectively. When most people speak of patents they are generally referring to utility patents, which account for over 90 percent of the patents issued.

How Long Does a Patent Last?

This depends on a number of factors. For example, design patents are granted for fourteen years, while plant and utility patents are granted for seventeen years. However, in 1980 renewal or maintenance fees were instituted for utility patents subsequently issued. These renewal fees are payable three and a half, seven and a half, and eleven and a half years after the patent is granted. If these are not paid within a stipulated period from the due date, the patent lapses. (See appendixes 4 and 5 for the most common fees.)

What Is a Utility Patent?

A utility patent protects the *functional* characteristics of any machine, manufactured product, process, or composition of matter, and is in force for seventeen years. As of January 1988 there were 4,500,000 utility patents. Historical examples of utility patents include U.S. patent 1,721,530, granted to Jacob Schick for "Shaving Implement" (electric razor); U.S. 621,195, granted to Ferdinand Graf Zeppelin for the "Navigable Balloon"; and U.S. 2,956,114, granted to Charles P. Ginsburg, Shelby F. Henderson, Ray Dolby, and Charles E. Anderson, for "Broad Band Magnetic Tape System and Method" (videotape recorder).

It is interesting to note that the numbering system for utility patents did not begin until 1836. All patents issued before that time are known as *name date* patents. From 1793 until 1836 approximately 600 patents were issued each year. These can be found by name or date in the *U.S. Serial Set* (U.S. Congress 1789-).

Why Would Anyone Need Access to Early Patents?

Early patents may have historical interest, such as U.S. 157,124, issued on November 11, 1874, to Jos. Glidden for "Wire Fences"; this is the first barbed wire patent issued. Another example is U.S. 6281, granted on April 10, 1849, to Walter Hunt for "Pin." Prior to this date there were no safety pins. This is the patent for this common and valuable device.

These are examples of important or interesting historical patents, but there are other reasons for searching prior art patents.* First and foremost is that early patents in another art area may be relevant to current technology; for example, some buggy whip patents foreshadowed the development of flexible "whip" car radio antennas.

How Is *Machine* Defined?

A *machine* is generally defined as a combination of heterogeneous mechanical parts adapted to receive energy and transmit it to an object.

Example: U.S. patent 1,103,503, granted to Robert H. Goddard for "Rocket Apparatus" (rocket engine).

Example: U.S. patent 2,790,362, granted to Louis M. Moyroud and Rene A. Higonnet for a "Photo Composing Machine."

Example: U.S. patent 4,841,832, granted to Gary L. Snavely and Richard Pelachyk for a "Boltrunner."

---

*See chapter 4 for a discussion of *prior art*, or the glossary for a definition of this important concept.

How Is a *Manufactured Product* Defined?

Manufactured is defined quite broadly as anything that industry can produce that is not a machine or composition of matter. A clearer description might be "article of manufacture."

Example: U.S. patent 775,134, granted to King C. Gillette for "Razor" (safety razor).

Example: U.S. patent 4,741,570, granted to Ann B. Lovass for a "Vehicle Bed Cover Assembly."

How Is a *Process* Defined?

A *process* is defined as a mechanical, electrical, or chemical procedure that involves one or more steps in the manipulation or treatment of some physical thing.

Example: U.S. patent 1,049,667, granted to William M. Burton for the "Manufacture of Gasoline."

Example: U.S. patent 4,842,575, granted to Harmon Hoffman and Kemal Schankereli for "Method for Forming Impregnated Synthetic Vascular Grafts."

What Is *Composition of Matter*?

Composition of matter is an intermixture of two or more ingredients, where there is more to the mixture than the mere effect of its individual components. It must be more than the type of mixture (recipe) found in a cookbook; lasagna and brownies are not compositions of matter, but drugs and alloys are.

Example: U.S. patent 942,809, granted to Leo H. Baekeland for "Condensation Product and Method of Making Same" (Bakelite™).

Example: U.S. patent 2,230,654, granted to Roy J. Plunket for "Tetrafluoroethylene Polymers."

Example: U.S. patent 2,485,760, granted to James M. Sprague for "Diazine Compounds (Sulfonamide)."

What Is a *Design Patent*?

Design patents protect new and original *ornamental* design of manufactured products. The design must be nonfuctional (otherwise it would be a utility patent), and integrally part of the manufactured product. The design patent protects only the appearance of an article, and not its structure or utilitarian features. Protection is limited to fourteen years. There are currently about 305,000 design patents. Design patents have their own sequential numbering system.

Examples: U.S. design patent 284,079, granted to Toshio Kita and Masfuku Akatsu for a "Telephone Instrument"; or design patent 11,023, granted to Auguste Bartholdi for the "Design for a Statue of Liberty."

What Is a *Plant Patent*?

Plant patents cover asexually reproduced living plants, such as roses or apples. An asexually propagated plant is one reproduced by means other than seeds, such as by the rooting of cuttings, by layering, budding, grafting, etc. Protection is granted to the entire plant. This protection has been granted to plants since 1930. By January 1990 approximately 7,000 plant patents had been issued.

The area of plant protection causes a bit of confusion because protection for sexually reproduced plants is a function of the Department of Agriculture, while protection for asexually reproduced plants is in the Patent Office, Department of Commerce.

Example: U.S. plant patent 1, granted to Henry F. Bosenberg for "Climbing or Trailing Rose."

Example: U.S. plant patent 3,222, granted to Richard A. Hensz for the "Grapefruit Tree."

Example: U.S. plant patent 6,878, granted to Hermann Finger for "Kalanchoe Plant Named Fantasy."

What Cannot Be Patented?

Ideas, results, functions, methods of doing business, laws of nature, or scientific principles cannot be patented. "Rational man," mathematical symbols and functions, "management by objective (MBO)," "first-in-first-out (FIFO) depreciation," gold, and gravity are all examples of things that cannot be patented. There are also some situations in which an inventor cannot secure a patent. These are referred to as the "statutory bars," and are as follows:

1. Knowledge or use by others in this country before the applicant's invention.
2. Invention patented or described in a printed publication in this or a foreign country.
3. Invention in public use or on sale in this country more than one year prior to the date of the application.
4. Invention patented or described in a printed publication in this or a foreign country more than one year prior to the date of the application.
5. Abandonment of the invention.
6. Invention patented on a foreign application filed more than twelve months before the U.S. filing.
7. Applicant is not the first inventor.

Summing up patentability, to be patentable an application must be a process, a machine, an article of manufacture, or a composition of matter; it must be new, useful, and avoid all the statutory bars. Definitions for *new* and *useful* are found in chapter 4, "Organization of the Patent File," and in the glossary.

Does an Invention Need to Work to Get a Patent?

The best answer to this is an example, and the drawing speaks louder than words. Patent 4,666,425 was granted May 19, 1987, to Chet Fleming for "Device for Perfusing an Animal Head" (see figure 2.1).

Claim 1 states that this is "a device for maintaining metabolic activity in a mammalian head which has been severed from its body at its neck."

Many inventions do not "work" or have not worked well because they were before their time. A good example is U.S. patent 392,046, granted in 1888 to John J. Loud for the "Pen" (ballpoint pen). This invention really took off after World War II with the invention of adhesive ink. Loud's pen did not work well because the ink he had at his disposal was runny and would not stick to either the ball or the paper. Even though his invention was not a commercial success he did reduce his idea to practice (capable of performing, not against laws of nature) and as a result, he is the inventor of the ballpoint pen. Basically an invention must be "capable of performing"; it does not actually need to perform.

Fig. 2.1. Patent 4,666,425, "Device for Perfusing an Animal Head." From *OG*, May 19, 1987.

4,666,425
DEVICE FOR PERFUSING AN ANIMAL HEAD
Chet Fleming, St. Louis, Mo., assignor to The Dis Corporation, St. Louis, Mo.
Filed Dec. 17, 1985, Ser. No. 809,949
Int. Cl.⁴ A61M *37/00*
U.S. Cl. 604—4 **20 Claims**

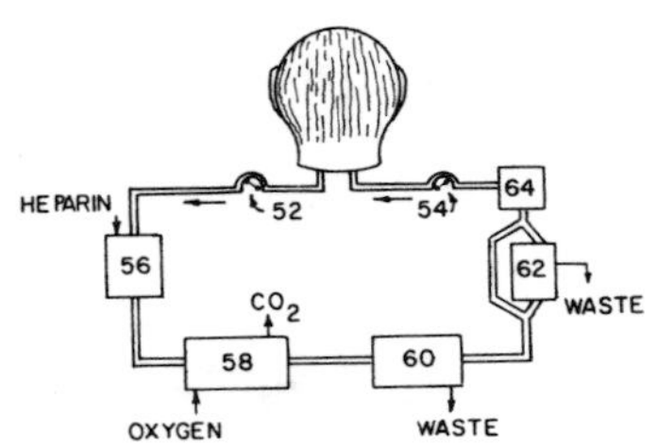

1. A device for maintaining metabolic activity in a mammalian head which has been severed from its body at its neck, comprising the following components:
   a. veinous cannulae which are capable of being attached to veins which pass through the neck and receiving blood from the veins;
   b. arterial cannulae which are capable of being attached to arteries which pass through the neck and transporting blood into the arteries;
   c. an oxygenation device which is in fluid communication with the veinous and arterial cannulae, and which is capable of displacing carbon dioxide contained in the blood with oxygen;
   d. one or more pumps of a selected type which causes relatively low levels of damage to blood components;
   e. fluid conduits which are attached to each of the components listed above in a manner such that the components, when coupled to the veins and arteries of a severed head by means of the cannulae of parts (a) and (b), will form a system capable of circulating blood through the oxygenation device and through the head after the head has been severed from the body; and,
   f. means for securely mounting the head upon the device after the head has been severed from the body, in a position such that the veins and arteries which emerge from the head can be connected to the veinous and arterial cannulae.

How Does One Get Protection for an Idea?

Ideas are the tools of the inventor. While every invention is based on at least one idea, every idea does not result in an invention. To get a patent, the idea must be "reduced-to-practice" through a complete description of the invention, its parts, and how it works. For example, suppose an inventor has the idea to design/make a faucet that protects against the possibility of being scalded by hot water coming unmixed or directly from the water heater. In order to get a patent the inventor would have to describe exactly how this idea would work, including drawings of the invention. For example, does this invention involve some type of thermostat? a valve? a combination? Is it an add on, or integral with the faucet? The patent application must answer these questions and others if the office is to grant a patent.

A useful analogy may be the example of an idea for a movie. First, after some thought, the author would send out a proposal for a movie to producers, etc. This is often referred to as a concept or treatment. This proposal gives an overview of the idea and might suggest a director or actors, but it does not include dialogue, stage directions, camera angles, and so on. It is the representation of an idea. In this analogy a movie concept is the same as an inventor's idea, and the screenplay is analogous to the inventor's patent application. In other words, in order to get a patent on an idea, it must be fully described, just as a screenplay must have the staging directions and the actors' dialogue.

Can an Invention Have More Than One Patent?

Only one patent is permitted per invention, and the same invention cannot be claimed in more than one patent. However, it is possible to patent improvements on existing inventions, aspects previously unpatented, or machines for making other inventions. For example, it is possible to have a patent on a device (for example, an egg timer) while at the same time also having a patent on the machine that manufactures the device. It is also possible to have a patent for a specific egg timer mechanism and to have a design patent for the novel "look" of the same egg timer. These are not patents on the same invention, but rather on different aspects of the invention.

Can an Invention Have Both a Patent and a Trademark?

Yes, because a trademark is any word, name, symbol, configuration, device, or any combination thereof used to identify or distinguish goods and services from those of others in the marketplace. Remember that both utility and design patents are generically called patents.

Example: G. D. Searle & Company has both a trademark (Nutrasweet) and a patent (3,800,046, "Artificially Sweetened Consumable Products") on the same sweetener. In this example, it would also be possible for Searle to have a design patent on the look of the capsule or tablet.

Can an Invention Have Both a Copyright and Patent?

No, copyright protects against the literal copying or reproduction of text, it does not protect ideas or inventions. However, it is possible for a device to be patented and the manual or operating instructions to be copyrighted.

What Is a Typical Patent?

There are no typical patents, just as there are no typical inventors. Some patents represent the product of an individual and some the combined efforts of a research team. Some are remarkably simple while others are very complex.

What Is an Invention?

There is no single definition of an invention, although over the years the courts have formulated a series of negative rules. These are expressed as generalities, and numerous exceptions have occurred. However, generally speaking, to be patentable an invention must be new, novel, or unobvious. Consequently, if an invention meets one or more of the following conditions, it is *not* considered to be new, novel, or unobvious, and is therefore unpatentable.

1. Substitution of a superior for an inferior material does not amount to invention (for example, gold for lead).

2. A change of size in a machine is not ordinarily an invention. Therefore, a smaller version of an existing pump is probably not a new invention.

3. A change of form or shape is not normally an invention.

4. Mere provision of adjustability does not amount to invention. For example, making an existing chair adjustable is considered an obvious invention and is not patentable.

5. Diminution of parts: for example, reducing the size of a computer chip alone does not make the new chip patentable.

6. Omission of a part and its function: for example, removing the gouging end on a swivel potato peeler is not a novel invention.

7. Use of old art or process for an unusual purpose does not constitute an invention. It is not a new invention to use a cigarette lighter to light a gas grill.

8. Mere aggregation is not an invention. This refers to a collection of parts or devices: for example, a recipe for tortillas is an aggregation, not an invention.

What Economic Value Does a Patent Have for Its Inventor/Licensee?

Monetary value is determined in the marketplace; the patent's value is determined by what a buyer will pay for it.

Is It Possible to Get a Copy of a U.S. Application?

"No information will be given by the Office respecting the filing by any particular person of an application ... without written authority" (37 *CFR* 1.14). Generally, the only applications that are not kept confidential are those of the U.S. government. (For more detail, see appendix 6. See also appendix 7.) The primary reason for this secrecy is that the United States is a "first to invent" country, rather than "first to file." In the United States the first person who can demonstrate that he or she reduced an invention to practice is granted the patent, and this necessitates that applications be kept secret until the patent is issued. On the other hand, in first to file countries, such as the United Kingdom, Germany, or the Netherlands, applications are made public because the inventor who demonstrates that he filed first is granted the patent. However, it is sometimes possible to discover the general outline of a U.S. application by finding a "patent equivalent." This is only possible when the inventor has been granted protection in another country and has listed his or her U.S. application number in that filing. More information on searching patent families or equivalents is found in chapter 10, "Online Patent Information Systems."

Are Patents Ever Granted in Secret?

All applications filed in the Patent and Trademark Office are screened for subject matter that might have an impact on national security. If the office concludes that disclosure would be detrimental, the Commissioner is notified, and he then issues a secrecy order that withholds the grant for a specified period (35 U.S.C. 181). For example, in 1944 and 1945 a series of applications were filed which, ten years later, resulted in U.S. patents 2,708,656 and 2,735,811 being granted to Enrico Fermi, Leo Szilard, Alvin Weinberg, Gale Young, Philip Morrison, and Leo Ohlinger for the technology which made the original atom bomb possible.

How Does One Get International Patent Protection?

The inventor must apply for a patent in every country where he or she wants protection, because a U.S. patent grants protection only in the territory of the United States. See chapter 11, "Foreign Patents," for more information.

What Is Contained in a Patent Application?

First, it should be noted that there is no patent application form for the inventor to fill out. Instead, when filing an application, the inventor or his or her agent must include the following:

Self-addressed receipt postcard

Transmittal letter

Check for the filing fee

Drawing or drawings of the invention

Specification containing the following:

- name of the invention
- background of the invention
- description of the drawings
- description (disclosure) of the structure of the invention
- explanation of how the invention works
- claims
- abstract

Patent application declaration form (sworn oath that application is true); see appendix 8.

Small entity declaration, if the inventor is an individual and has not transferred rights to a large entity (for-profit company with more than 500 employees); see appendix 9.

Can an Employer Obtain a Patent on an Employee's Invention?

No, only the actual inventor can patent an invention. An employer, however, can own the rights to an invention. For the purposes of patenting, inventors must be persons. Corporations, partnerships, or associations, because they lack the ability to discover or invent, cannot be listed as inventors, but they can be listed as assignees. The fact that an employer paid for the invention has no impact on who is listed as the inventor.

Does One Have to Submit a Model of the Invention?

Models were once required for all patent applications. Now a model will not be accepted unless specifically requested by an examiner. Models or specimens are most commonly required for alleged perpetual motion machines, composition of matter, and microbiological inventions. When the invention relates to a composition of matter, the applicant may be required to furnish specimens for inspection or experiment. For microbiological inventions, a deposit of the organism is required.

What Do "Patent Applied For" and "Patent Pending" Mean?

These phrases are often used by a manufacturer or seller of an article to inform the public that the inventor has filed an application for a patent. The law imposes a fine for falsely using these terms.

How Long Can a Patent Be "Pending"?

Until the entire procedure is completed, including, if applicable, completion of office actions, any interference procedures, appeal to the Patent Board of Appeals, and appeal to the U.S. Court of Appeals for the Federal Circuit for decisions. Decisions can be "all claims rejected," "some claims allowed," "remand," and, rarely, "reopening of prosecution." An example is the ENIAC patent first filed June 26, 1947, which finally was issued as U.S. patent 3,120,606 on February 4, 1964. During the eighteen-year pendency the applicants were involved in eleven interferences.

What Is an Interference?

An interference is a proceeding instituted for the purpose of determining the question of priority of invention between two or more inventors (37 C.F.R. 1.202).

What Are Statutory Invention Registrations?

These registrations of inventions are printed and published like patents but have no enforceable rights since they are not patents. Once issued, they preclude others from obtaining a patent on the disclosed invention and are generally considered to be a defensive procedure.

Examples: H419, "Contoured Punch Tool for Removing Semi-Tubular Rivets," issued to Halver V. Ross, Bountiful, Utah, assignor to the United States of America as represented by the secretary of the Air Force, Washington, D.C.

H758, "Sealant Composition," issued to Steven S. Chin, Houston, Texas.

H760, "Silver Halide Photographic Materials," issued to Tadashi Ogawa, Kanagawa, Japan, assignor to Fuji Photo Film.

For a complete example of a statutory invention registration, see appendix 10. Statutory invention registrations are published in the *Official Gazette* in a separate section, arranged by sequential numbers preceded by an *H*. More specific information on the rights associated with statutory invention registrations can be found in the United States Code (35 U.S.C. 157).

Reasons for obtaining one of these are varied, but may include (1) expense—the application fee is less than for patents, and there are no post issuance fees; (2) a company wanting credit, but having no expectation that the idea is worth pursuing; or (3) an individual wishing to be "published." In other words, the reasons are as individual as the inventors themselves.

What Does It Mean When a Patent Is Reissued?

The *Manual of Patent Examining Procedure* (*MPEP* 1989), Chapter 1400 "Correction of Patents," states that "a reissue application is an application for a patent to take the place of an unexpired patent that's defective in some one or more particulars" (*MPEP* 1989). The most common defects include claims that are too narrow or too broad; a disclosure that contains inaccuracies; failing to claim, or incorrectly claiming foreign priority; and failing to make reference to or incorrectly making reference to prior co-pending applications. (See chapter 14 for more information on the *MPEP*, and appendix 11 for sample pages.)

Reissues are listed in the *Official Gazette*, and a consolidated list is available on microfilm in patent depository libraries and from Research Publications, Inc. An example of a reissue can be found in appendix 12.

What Do "Continuation Application" and "Continuation-in-Part" in *Official Gazette* Entries Mean?

"A continuation is a second application for the same invention claimed in a prior application and filed before the original becomes abandoned" (*MPEP* 1989, 201.08). A continuation-in-part is "an application filed during the lifetime of an earlier application by the same applicant, repeating some substantial portion or all of the earlier application and adding matter not disclosed in the earlier application" (*MPEP* 1989, 201.08).

Is It Possible to Discover What Patents Have Been Abandoned, Had an Adverse Decision or Been Reissued?

Information on reissues, adverse decisions, and so forth is available at the Patent and Trademark Office and in patent depository libraries in the "CDR File." The CDR File has two parts, an annual cumulative index and CDR microfilm. The cumulative index is arranged by patent number and provides access to the reel and frame numbers of the CDR microfilm. The microfilm consists of copies of any reissued patents, disclaimers, and reexamination certifications that are associated with the original patent number. A similar publication, the *Patent Status File*, is available from Research Publications, Inc., and covers corrections from 1973 to 1988. Several commercial databases can also be searched for status information. (See chapter 10.)

Is It Possible to Discover Which Patents Have Expired for Failure to Pay Maintenance Fees?

Maintenance fees are required for utility patents, but not for design or plant patents. A list of expired utility patents is printed in the front of the *Official Gazette*. In accordance with 35 U.S.C. 41 and 36 CFR 1.363(g), patents expire at the end of the fourth, eighth, or twelfth anniversary of the granted patent, depending on the first maintenance fee not paid. Maintenance fees are required for all original or reissued patents filed on or after December 12, 1980.

Status, such as expiration for failure to pay maintenance fees or whether a patent has been withdrawn, can also be found on CASSIS CD-ROM* for utility patents issued from 1969 to the present, and for all others since 1977.

What Happens to a Patent When It Expires?

After a patent expires, both disclosed and claimed technology are in the public domain, and anyone can make, sell, or use the invention without the patent owner's permission as long as the patent rights of someone else are not infringed upon. Therefore, all legal rights cease when a patent expires, including the right to license the invention and the ability to collect royalty payments.

Can One Always Get a Copy of Any U.S. Patent?

U.S. patents are always in print. However, a few early patents were lost in a fire at the Patent Office. For these the only record is in the *U.S. Serial Set, Report of the Commissioner of Patents*. (See appendix 14.) Note that these are the documents a searcher uses to find out what was issued prior to the publication of the first *Official Gazette* in 1872. In any case, other than these few, U.S. patents are always kept in print.

The next chapter focuses on the patent as a document by describing its structure and format, defining the parts of a patent, and describing the bibliographic elements found on the first page of all modern U.S. patents.

REFERENCES

Code of Federal Regulations, 2d ed. 1949- . 1.14 Title 37 CFR. Washington, D.C.: Superintendent of Documents.

______. 2d ed. 1949- . 1.202 Title 37 CFR. Washington, D.C.: Superintendent of Documents.

*CASSIS CD-ROM is available in all patent depository libraries (PDLs). For a list of all PDLs, see appendix 13.

Manual of Patent Examining Procedure (MPEP). 1989. Chapter 1400 "Correction of Patents." Washington, D.C.: U.S. Patent and Trademark Office.

______. 1989. Chapter 201.08 "Continuation-in-Part Application." Washington, D.C.: U.S. Patent and Trademark Office.

United States Code. 1982. 35 U.S.C. 181. Washington, D.C.: Government Printing Office.

United States Congress. 1789-1969. Congressional Series of United States Public Documents, Congressional Set. Washington, D.C.: Government Printing Office.

ADDITIONAL READINGS

Brown, E. W. 1986. "Patent Basics: History, Background, and Searching Fundamentals." *Government Information Quarterly* 3, 4: 381-405.

Harmon, R. L. 1988. *Patents and the Federal Circuit*. Washington, D.C.: Bureau of National Affairs.

Ojala, M. 1989. "A Patently Obvious Source for Competitor Intelligence: The Patent Literature." *Database* (August): 43-49.

Rosenberg, P. D. 1980. *Patent Law Fundamentals*. New York: Clark Boardman.

"Special Issue on Patents." *IEEE Transactions on Professional Communications*. 1979. PC-22, 2: 46-127.

Tertell, S. M. 1986. "Patents Are an Overlooked Information Source." *Bulletin of ASIS* 13, 1 (October/November): 24-25.

U.S. Department of Commerce. 1987. *Basic Facts about Patents*. Washington, D.C.: Government Printing Office.

______. 1985. *General Information Concerning Patents*. Washington, D.C.: Government Printing Office. 003-004-00583-1.

______. 1966. *Development and Use of Patent Classification Systems*. Washington, D.C.: Government Printing Office. C21.14/2:C56.

3 The Patent As a Document

The Congress shall have Power ... to promote the Progress of Science and Useful Arts, by securing for limited Times to Authors and Inventors exclusive right to their respective Writings and Discoveries.

—United States Constitution, Article I, Section 8

This chapter describes in detail the parts of a patent, so that any researcher will be able to examine a patent, quickly finding the section he or she needs. This information is important because a patent is a specialized government document that has both technical and legal aspects. As a technical document, the patent describes the invention and its functions so that anyone "skilled in the art," or competent in the same field, can reproduce the invention. As a legal document, the patent sets forth the fundamental aspects, such as "newness" or "novelness," of the invention and states or claims that these are original to the inventor. With the issuance of the document, the inventor is granted a limited monopoly to "exclude others from using the invention or selling the invention" (Taft 1928).

STANDARDIZATION

Over the years U.S. patents have been issued in a number of different physical formats. The current format (see appendix 1) has been used since August 1970 and has a well-defined structure and format (discussed in this chapter). While the format and type of patent issued vary from country to country, all patents contain certain bibliographic data. Fortunately, procedures for standardizing these common bibliographic elements have been agreed upon in the Patent Cooperation Treaty (PCT) and by the World Intellectual Property Organization (WIPO), both of which agreements have brought a high level of order and standardization to the patent documents issued all over the world.

In the United States, the WIPO standard, St. 9 (PCPI/GI/IV/3), was slightly modified by the American National Standards Institute (ANSI) and issued as ANSI standard Z39.46-1983 to reflect U.S. practice (WIPO 1983; ANSI 1983). Both the ANSI and WIPO standards reflect the minimum data that should be printed on the first page of a patent document or published as an entry in an official gazette or journal.

As a result, users can identify bibliographic data without knowledge of the language used because each data element is defined and assigned a number. See appendix 15 for an example of a patent from the People's Republic of China. These "Internationally agreed Numbers for Identification of Data," or INID numbers, can be found in brackets on the first page or cover sheet of a U.S. patent. The definitions of these bibliographic elements (fields) are listed below. The definitions are given in numerical order, not in the order in which these elements appear on the first page of a U.S. patent. These field numbers are used on all three types of patents issued in the United States. INID numbers are broken down by category and then further subdivided by number field name, as follows:

Document Indentification [10]:

[11] Number of the document

[19] Country code or the plain language indentification of the country publishing the document

Domestic Filing Data [20]:

[21] Number(s) assigned to the application(s)

[22] Date(s) of filing application(s)

[25] Date of publication by printing on which grant has taken place

Technical Information [50]:

[51] International patent classification

[52] U.S. classification

[54] Title of the invention

[56] List of prior-art documents

[57] Abstract

[58] Field of search

Identification of Parties Concerned [70]:

[73] Assignee/grantee

[75] Name(s) of inventor(s) who is(are) also applicant(s)

[76] Name(s) of the inventor(s) who is(are) also applicant(s) and grantee(s)

Fields not used in the United States but used in other countries include [30] Foreign Priority Application Data, [60] References to Other Legally Related Domestic Patent Documents Including Unpublished Applications, and [80] Identification of Data Related to International Conventions Other Than the Paris Convention.

The remainder or text of a U.S. utility patent has no INID numbers, but consists of various elements that are referred to generically as "the specification." For ease and convenience in locating information in the specification, the columns and lines are numbered.

ELEMENTS OF U.S. PATENTS

The following discussion describes and defines the basic parts common to all U.S. design, plant, and utility patents. A short discussion of the elements of patents that are not common to all patents follows. These differences occur because plant and design patents protect inventions that differ substantially from those granted utility protection. The differences are recognized by separate numbering systems and by different formats.

The specification elements common to all patents include:

Title: A brief restatement of the title reflecting the main element of the invention.

Abstract: A brief synopsis or summary of the invention.

Background section: A brief statement concerning the general and specific field or area of the invention, including cross-references to related applications; relevant prior art; a list of related applications filed by the inventor, if relevant; and a discussion of prior art or how the invention related to previous attempts to solve the problem.

Objectives and advantages section: A summary of the invention. The term *objectives* refers to "what the invention accomplishes" and should include a discussion of the advantages of the invention over previous inventions. The description of the invention, also known as the preferred embodiment, should disclose how the invention works and should include descriptions of the static features of the invention, part-to-part interconnections, any exotic parts or materials, and the process or machinery involved in the invention. The description must set forth in clear, precise language the best way, according to the inventor, of carrying out the invention.

When the invention relates to an improvement in a process, machine, article of manufacture, or composition of matter, the specification must indicate the part or parts that were improved.

Description of drawings: A series of separate paragraphs describing the drawings. The descriptions must include extensive detail on the operation or function of every part of the invention along with a description of how each part works and its relation to other parts. Each description is numbered and relates to the same number on the drawing. The descriptions and the drawings are important, because the drawings are often relied on to explain details of the invention.

The applicant for a patent is required by law to furnish a drawing of the invention whenever possible or relevant. This includes practically all inventions except compositions of matter. Because they do not have drawings, chemical patents (composition of matter) generally have detailed descriptions of the methods used in making the compound or compositions being claimed. In a chemical patent, the description should include starting materials or ingredients, preparations, any reactions, and procedures.

Drawings: Drawings which must show every feature of the invention specified in the claims and must contain as many figures and views as necessary to show the invention. The Patent and Trademark Office specifies the size of the sheet on which the drawings are made, the type of paper, the margins, and other details. The reason for specifying these details is that the drawings are printed and published in a uniform style when the patent is issued, and the drawings must be readily understood by persons using the patent descriptions. The rules for drawings are found in the Code of Federal Regulations (see appendix 16).

Claims: Claims recite and define the structure of an invention in precise, logical terms. They point out and distinctly claim the subject matter that the applicant regards as his or her invention. Therefore, the purpose of the claims is to discuss and delineate the scope of legal protection given the invention. This means that the concepts of novelty and patentability are judged by the claims, not the disclosure. Consequently the claims are the operative part of each patent.

Every claim can be classified into one of five statutory classes of invention: (1) process or method, (2) machine, (3) article or article of manufacture, (4) composition of matter, and (5) those claims showing new usage of previous classes of invention. Each claim can be independent and stand alone, or the claim can be dependent and refer back to a previous claim.

Claims should be written in a narrow enough fashion to differentiate the invention from others, but broad enough to ensure protection for the inventor. They must conform to the invention as set forth in the remainder of the specification, and the terms and phrases used must find clear support in the description so that the meaning of the claims may be ascertained by reference to the description. Claims are the essence of the invention; questions of infringement are judged by the courts on the basis of the claims. The importance of the claims cannot be overstated; this section defines the property rights of the patentee. See appendix 1 for an example of a utility patent specification.

Design Patent Specification

Like utility patents, design patents have all the standard bibliographic elements, except that they have no abstract. An abstract is not necessary because a design patent provides protection only for the "look of" the protected subject. The description section does exist, but consists entirely of references to the drawing. Probably the most significant difference is that only one claim is permitted. This claim must refer to the article designed by its title and must make reference to the drawings. The drawings of a design patent follow the same rules as drawings in utility patents.

Obviously then, the drawings in a design patent are very important and provide the best description of the item. The term *design* is somewhat misleading, since the patent protects the aesthetic attributes of the article, not its functional attributes. See appendix 2 for an example of a design specification.

Plant Patent Specification

Like utility and design patents, plant patents have the same standard bibliographic descriptions. In a plant patent the specification describes the plant and the characteristics that distinguish it from other related known plant varieties and antecedents in standard botanical terminology rather than using the broad nonbotanical characterizations commonly found in nursery or seed catalogs. The specification should also include the origin or parentage of the plant variety and must point out where and in what manner the variety has been asexually reproduced. Where color is a distinctive feature the color should be positively identified in the specification by reference to a designated color in a recognized color dictionary.

Each plant patent consists of a single claim because the patent is granted to the whole plant. Like design patents, the drawings for plant patents are very important and even more specialized than either design or utility patents. Plant drawings must be in scale, be botanically correct, show the vegetative/fruiting parts, and should be artistically and competently executed. See appendix 3 for an example of a plant patent specification.

Design, plant, and utility patents are often written in highly stylized technical legal language and are issued in a controlled format. Consequently it is important to know how to examine and recognize the parts of a patent, so that the type of information being sought can be located by looking in the correct part of the patent. For example, when conducting an actual patentability search, searchers will need to read and study each patent retrieved in order to discover if their invention has been patented. This will mean closely examining the claims. If, on the other hand, the searcher is looking for information or a solution to a particular problem, he or she would want to read closely the technical disclosure, which consists of the background and the objectives and advantages sections. If the examined patent is closely or highly related to what the searcher is looking for, he or she might be wise to look at the cross-references.

The next chapter discusses how the patent file is organized, describing how patents are classified and giving a brief history of the development of the classification system. This classification system provides the underlying organization for the patent file and currently consists of the approximately 5,000,000 issued patents. It is this system which brings together all related patents and allows searching by the subject of invention. Understanding this classification system facilitates searching and increases retrieval accuracy; the importance of the classification system for searching U.S. patents is enormous.

REFERENCES

American National Standards Institute (ANSI). 1983. *Patent Documents—Identification*. Z39.46. New York: American National Standards Institute.

Taft, W. H. (Chief Justice). 1928. United States vs. General Electric. 272 U.S. 485.

World Intellectual Property Organization (WIPO). 1983. *Identification of Patent Bibliographic Data*. Lanham, Md.: UNIPUB.

ADDITIONAL READINGS

Cottone, J. R. 1979. "Writing and Invention Disclosure." *IEEE Transactions on Professional Communications* PC-22, 2: 105-8.

Flanagan, J. R. 1983. *How to Prepare Patent Applications*. Troy, Ohio: Patent Educational Publications.

Greer, T. J. 1979. *Writing and Understanding U.S. Patent Claims*. New York: Michie Bobbs-Merrill.

"Guide for Patent Drawing." 1979. *IEEE Transactions on Professional Communications* PC-22, 2 (June): 109-11.

Maynard, J. T. 1979. "How to Read a Patent." *IEEE Transactions on Professional Communications* PC-22, 2 (June): 112-18.

4 Organization of the Patent File

A man has no ears for that to which experience has given him no access.
—Friedrich Nietzsche

Described here are the purpose and elements of the U.S. patent classification system. A summary of the principles on which this system is based is also provided. Without an understanding of how patents are organized (classified) into a searchable and retrievable file, the searcher will not be able to use the classification to its best advantage and will at best perform a misleading search. By not using the classification system correctly the searcher could miss relevant patents and waste valuable time and money. He or she might not find the information being sought or might apply for a patent on an already patented invention. The importance of the classification system for retrieval of information stored in U.S. patents cannot be overstated.

This chapter includes a short history of the classification system, an introduction to the patent classification system, the basis or rules of classification, and a hypothetical example with instructions describing how to use and read the classification system.

One of the hallmarks of the U.S. patent system is its highly controlled and structured organization. It would be impossible to find, much less group together, related subject matter in a system this large were it not for the development of a specialized indexing and filing system. The method used to accomplish this is systematic arrangement or subdivision of subject matter and is the major organizational technique used to organize patents. The classification system consists of an arrangement of all scientific and technical information (prior art) to facilitate the selective retrieval of specific information when desired. The prior art comprises both claimed inventions as well as technical disclosures in all domestic patents and all other publications. An understanding of how the classification system works is of practical interest to anyone attempting to use the patent file, for it is the classification system that allows users to locate and retrieve pertinent patents in an efficient manner.

INTRODUCTION TO THE U.S. CLASSIFICATION SYSTEM

This hierarchical subject classification system for U.S. patents consists of over 400 classes and 100,000 subclasses. Within the U.S. patent system similar subject matter is brought together in large groupings called classes. Each of these classes and subclasses has a unique number, title, and definition. New classes and subclasses are added, changed, and deleted continuously, so that the classes reflect, as accurately as possible, the technologies found in issued patents. The idea underlying classification is to create a flexible system, one that allows historical inventions to be classed with their more recent relatives and that can be expanded to allow for entirely new categories of invention. An example of these changes is in appendix 17, which compares pages from a modern schedule with pages from a 1923 schedule. Not only is the new schedule larger because of changing technology but it also allows for more specificity. This example also demonstrates keeping related historical patents together. Subclass 121, "Darning" includes all patents on darning regardless of when they were issued. An example of adding new classes to reflect changing technology is class 800, "Multicellular Living Organisms and Unmodified Parts thereof," which was created in 1986 to reflect the patenting of genetically altered animals. Located in class 800, subclass 1, is patent 4,736,866, "Transgenic Non-Human Mammals," the first patent in history to cover an animal.

Because the classification system is dynamic, it is important to remember that the classification number given in the *Official Gazette* or on the face of a patent was true at the time of issuance and may later be changed at any time. Finding the current classification is covered in chapter 10, "Online Patent Information Systems."

A fundamental principle of the classification system is that each class or subclass is developed by *first* analyzing the claimed disclosures of U.S. patents, *and then* creating the various divisions and subdivisions on the basis of this analysis. When placing a patent into the classification system, the examiner or classifier considers each claim separately and then assigns one or more classifications as required. Among the concepts considered when classifying an application are utility, proximate function, and structure. *Utility* means "usefulness when applied to a process or means" and refers to the function which is performed, the effect of the process or means, and the product which is produced by the process or means. It does not mean useful, as a broom is useful, but rather having a use for the invention. For example, a coffee mug with the handle inside the cup is not useful but is rather used *for* amusement. *Proximate* (nearest) *function* is an important concept that unfortunately eludes precise definition. Basically, the intent of proximate function is to bring together devices and structures that work in the same way to solve similar problems. For example, both knives and scissors use an incline plane or edge to cut and as a result they demonstrate proximate function. However, while both hammers and baseball bats can be used to "strike a blow," they do not solve the same problem and as a result do not have proximate function. *Structure* relates to subject matter that is so simple and of such general utility as to have no obvious functional characteristics. Its only distinguishing features are structural; for example, stock materials, such as metallic alloys; rubber laminate; sheet material for magnetic disks; or substrate material for semiconductor chips.

All of these important concepts are used to create a system that will provide storage and retrieval of all prior art (similar subject matter). The intent is to be exhaustive and to aggregate all patentable matter. This aggregated patentable matter is referred to as the prior art and consists of all pertinent and applicable public knowledge and experience known at the time a patent application is filed. Ideally, this system provides a reasonably short and complete search for each type of investigation being made, such as patentability, novelty, interference, and informational searches. This system gives searchers the reasonable assurance that the claimed subject matter contained in each patent is retrievable.

The U.S. patent classification system differs quite dramatically from other common classification systems, such as those used in libraries. In library classification systems, such as Dewey or the Library of Congress, the theoretical arrangement is determined first and each book is then placed within the existing system. When new subject matter arises, a new number is created; older related materials are rarely reclassed to the newly created number. This is directly opposite to the system used by the Patent and Trademark Office, which reclassifies all previous patents whenever a new number is created.

THEORY OF THE CLASSIFICATION SYSTEM

The classification system is designed to provide a way to clarify the relationships between similar devices and to associate related concepts, or those having a predetermined degree of likeness, into large groups. It divides these large groups into subgroups, and then arranges the subgroups into an orderly sequence that allows the user to determine the nature and significance of the subject matter and at the same time illuminates the relationship between the subgroups.

The classification system is in principle inclusive rather than exclusive. The class or subclass that provides for a particular concept also includes patents that claim the disclosed feature alone, the disclosed feature in combination with other features, and subcombinations or other features not specifically provided for elsewhere. For example, a class entitled "Metal Deforming" includes, unless specifically excluded, bending or forging, bending in combination with an assembly means, and work holders for use during bending. One reason for the inclusive nature of the classification system is that it is impossible to establish separate subclasses for an infinite number of possible combinations and subcombinations. This can be demonstrated by using metals as an example. Thirty metals can be combined to form 435 binary alloys, 4,060 ternary alloys, and 27,405 quaternary alloys, and this is without taking differences in proportion between the alloying metals into consideration. In fact, the total number of possible alloys of known metals is incomprehensible. A similar logarithmic increase can happen in patent classification.

Assignment to these all-important classes and subclasses is based on the patent's most comprehensive claim, rather than the title or abstract. Claims are used because they formally specify exactly what is *new* and therefore what is protected by the patent. The largest or most comprehensive claim is used to establish the original classification because it covers more of the disclosed technical matter and therefore acts as a summary of the patentable invention.

The claim section of a patent may start with the words: "I claim" or "What is claimed is." Claims generally contain words such as "comprising," "including," or "composed of." These words have a definite legal meaning. "Consisting of" and "composed of" mean that the patented invention is made up of the listed elements and no others. The terms "comprising" or "including" are much broader and require merely that the listed elements be present.

Placement of Patents in a Class

The initial or primary classification assignment is referred to as the "original" (OR) class number. When placing a patent into the U.S. patent classification system, the examiner or classifier considers each claim separately and then assigns one or more classifications as required by the *Manual of Patent Examining Procedure*. Many times a patent has several different claims, which if found in different patents would be classified in different subclasses. It is obligatory in such situations to cross-reference the patent to the subclasses that provide for the subject matter being claimed. These obligatory cross-references are referred to as "mandatory"; it is from these mandatory classifications that the original classification is selected. The remaining mandatory classifications are designated as cross-references (XR). Both the OR and XR numbers can be found on the first page of the patent in field [52]; the OR is listed first in boldface type and the XRs are next. Cross-references (XR) can also provide access to additional class numbers and therefore index other facets of subject matter not found or disclosed in the most comprehensive claim. Besides these mandatory cross-references, the classifiers and examiners, within certain guidelines, can designate any number of discretionary or additional cross-references. These are often used to provide access to subject matter not covered by the claims. Examiners and classifiers work in the same subject areas for years and as a result become extremely familiar with their assigned classes or "art group." In any particular patent application there may be interesting or useful information that appears in the disclosure section. Since mandatory classification is based on the claims, other disclosed information would not be cross-referenced were it not for these discretionary references. An examiner can decide to make a discretionary cross-reference to this disclosed information so that it can be retrieved at a later time. No distinction is made between mandatory and discretionary cross-references. As a result, the original and the cross-references should be considered as equals in terms of reference value. Any cross-reference numbers encountered by a search are pointed reminders that other subclasses may contain relevant subject matter. These cross-references are often referred to as "the field of search," and are found in field [58] on the front page of any issued patent.

It is important to be aware of a number of techniques that are used to limit the amount of cross-referencing. These techniques have important ramifications for searching and mean that the searcher should not assume that every patent has been cross-referenced to every pertinent class/subclass. The most important of these techniques follow:

1. The hierarchical arrangement of the subclasses in a class schedule precludes the need for cross-references.

2. Search notes located in the *Classification Definitions* (see appendix 18).

   Example: Class 800
   SEARCH THIS CLASS, SUBCLASS:
   862, for thin film type electrical element used for switching in a nonlinear solid state circuit.

3. Cross-referencing between classes is generally precluded by the existence of a search note in the definition.

   Example: Class 453 COIN HANDLING
   SEARCH CLASS:
   235, Registers, subclasses 7+ for cash registers which keep track of cash tendered.

4. Cross-referencing between subclasses is generally precluded when the relative schedule position of the subclasses indicates that a search would include both subclasses.

THE CLASSIFICATION SCHEDULE

The classification schedules are the physical representation of the hierarchical arrangement of the classes and subclasses. It is through these schedules that searchers access patents classed together by similar subject matter.

The best way to understand and appreciate the strengths and oddities of this classification system is through an example. The following is a hypothetical example that classifies scrap in a junkyard. The subject material of this example is scrap and the title of the hypothetical class is SCRAP. The subject material of scrap is divided into smaller and smaller subclasses, and the entire arrangement is called a class schedule.

Class title: SCRAP

1 COMBINED BAR, LINK AND BALL

2 COMBINED BAR AND LINK

3 COMBINED BAR AND BALL

4 COMBINED LINK AND BALL

5 CHAIN

21 .with end fastener

22 .with flaccid cover

23 ..removable

6 BAR

7 LINK

8 BALL

9 .hollow
10 ..perforated
11 ..grooved
12 .perforated
13 .grooved
14 .mineral
15 ..metallic
16 ...aluminum
17 ...zinc
18 .rubber
19 .ivory
20 .MISCELLANEOUS

This schedule is very similar to an outline used to write a term paper. Each indented heading (subclass) further qualifies the heading under which it is indented. In our hypothetical example subclass 21 is indented under subclass 5 and classifies chains with end fasteners. Each subclass must be read as including all of the limitations of the superior heading. For example, subclass 15 "metallic ball" scrap, will house *only* those items that are at least scrap metal balls. The use of the qualifier "only" is important and relates to the hierarchical nature of the classification scheme. Therefore, subclass 15 houses metallic ball scrap, and subclass 5 includes chains of unspecified type and material. Subclass 10 is more specific than 15 since it houses "perforated, hollow ball" scrap. In this example the balls called for in subclass 15 must be metal, while those in subclass 10 could be metal, plastic, or wood, as long as they are perforated hollow balls.

The use of capitalization and the placement of the dots is important to understanding the organization of the class schedule. Subclass titles (those capitalized) that do not have a preceding dot are referred to as "main line" or "first line" subclasses. In the example those numbered 1, 2, 3, 4, 5, 6, 7, 8, and 20 are main or first line subclasses.

Uncapitalized subclasses with dots are referred to as indented subclasses and are named by the level of indentation. For example, a subclass indented one level (or dot) below a main subclass is a "second line" subclass (e.g., 21, 22, 9); those indented two levels or dots are referred to as "third line" subclasses (e.g., 23, 10, 11). All subclasses indented under a superior concept are generally referred to as that concept's "indents." For example, numbers 16, 15, and 14 are all indents to subclass 8. In this example 16 is more specific than 15, which is more specific than 14. The complete title of subclass 16 should be read as: as aluminum metallic mineral ball scrap. This means that an item to be stored or classed into 16 must have all of the characteristics of subclass 16. As a result, iron ball scrap would be in 15 and quartz ball scrap would be in subclass 14. When reading the schedule it is important to read all of the indent levels as a sentence, checking to be sure that all elements are relevant to your search.

Subclasses positioned at the same level of indentation are referred to as "coordinate" subclasses, provided that some superior subclass is not located between the subclasses being compared. Consequently, subclasses 21 and 22 are coordinate with respect to each other, but subclasses 22 and 9 are not. This means that subclass 9 has *no* relationship with subclasses 22 and 21, even though they are all second line subclasses, because subclasses 6, 7, and 8 come between 22 and 21. However, all first line or main line subclasses are coordinate with respect to each other since nothing can be superior to a main line subclass. Therefore, 1, 2, and 7 are coordinate with each other and deal with the subject matter of link in different but equal or coordinate ways.

The order in which subclasses appear in a schedule establishes the order of superiority between the concepts provided for in the schedule. The actual number of the subclass has no function other than to provide an address or storage area for the patents. The numbers can get out of numerical sequence when a class is rearranged to reflect new technologies. This must have happened at one time in our example, since subclasses 21, 22, and 23 are now hierarchically superior and come before subclass 6. Again, it is the location of the subclass title within a schedule *hierarchy* that is important, not the number of the subclass.

Generally, subject matter is *hierarchically* arranged, with subclasses embracing the largest organization of elements (the most complicated structures) appearing higher in the schedule. Less complicated or simpler subject matter is found further down the schedule. Our example demonstrates this arrangement. We find that scrap consisting of a bar plus a link plus a ball (subclass 1) is the most complex. It is a higher assemblage of elements than a bar plus a link (subclass 2). Simpler items are found at the lowest level. In this example, the simplest scrap, ivory ball, is found in subclass 19.

A second important characteristic of the classification schedules is the *inclusive* or *exhaustive* nature of the subclasses. This means that a subclass will take *at least* the subject matter provided for in the subclass. Given the concepts of hierarchy and the inclusive nature of subclasses, the classification schedules can be viewed as a series of sieves or strainers located one above another. Those sieves located higher in the array have larger diameter openings, so that they strain out only the most complex (largest) elements claimed. As the sieves become smaller and smaller, they trap smaller sized "particles" of information. This process of narrowing and defining continues all the way down the schedule to the bottom.

If material is dropped into the top of a sieve array, it will pass through the array until an opening is reached that blocks further passage. This is how the modern classification schedule works. When we want to place or retrieve (search) a concept using a schedule, we start with the first subclass and then proceed to move down through the first line subclasses (those without dots) until we find one that will accept the subject matter. If this subclass exhausts the claimed subject matter and all concepts employing this subject matter, we should not have to look at any lower first line subclasses.

In the scrap example, suppose as we are sorting through our pile we run across an unusual style of bar. The next step would be to decide where to store the bar. We would consult our schedule, starting at the top and scanning down. Subclass 1 provides for bars, but these must be combined with links and balls. Since all three must be present, classification 1 is not appropriate. All we have is a bar, however unusual. Subclasses 2 and 3 also provide for bars, but again bars found here are combined with other features such as links. The first subclass that will accept a bar by itself is 6. Therefore, all bar scrap should be in subclass 6.

Now, let us suppose that after a few months, someone comes in looking for an unusual type of bar. We know from our previous experience that bars per se are in subclass 6 and since all subclasses are exhaustive, there is no need to look lower in the schedule than 6. But what about higher subclasses 1, 2, and 3, which all include bars as part of their organization? These subclasses must be searched if we want to be sure that we have seen all unusual bars, no matter what they are combined with. In other words, the unusual bar might be found in subclass 4 because the bar also included a ball. Subclass 6 includes plain bars, no matter how unusual.

Finally, once we have selected a subclass, we must investigate its indents and read the subclass definition and any notes to be sure that an exception has not been made. Exceptions do exist and care must be taken to avoid wasting time and effort.

The class schedules in current use by the Patent and Trademark Office are collected in numerical order in the *Manual of Classification* (see appendix 19). Unlike our hypothetical example, which was chosen to demonstrate location and indentations clearly, functioning class schedules have a number of other important designations that can facilitate use of the schedule, such as "digests," "cross-reference art collections," and "unofficial or alpha subclasses."

Digests and X-Art Collections

A digest is a collection of cross-reference copies of patents created by an examiner. They are "outside" of the official classification arrangement and consist of cross-referenced patents that relate to a class but not to any specific subclass within that class. They were created by the examiners over the years to facilitate searches. For example, class 123 digest 12 is a collection of hydrogen fueled engines brought together from several different subclasses in class 123, "Internal Combustion Engines." Digests are listed at the end of the schedule and are not defined. It should be noted that the creation of new digests is generally not allowed. Cross-reference art collections are now taking the place of digests.

Cross-reference art collections (X-art) are official collections of patents based on a concept other than proximate function. As a result, all patents in an X-art collection are cross-referenced into this X-art collection. These collections were created to facilitate searching by grouping together patents that, because of the rules, cannot be grouped together. Only cross-referenced (XR) patents are listed in an X-art collection. Unlike the old digest, X-art collections are both inventoried and defined. The value of these collections is demonstrated by the definition for class 505 X-art 800, which collects into one group the patents concerning the making of superconductive materials. Normally, these patents would be classified in many different classes and subclasses depending on the process used to make the material. This is clearly demonstrated by the following definition for class 505, "Superconductor Technology," X-art 800, "Superconductive Material, per se, process of making same": Art Collection under the class definition including (a) substance(s) used to make a superconductor, (b) process of making the substance(s), (c) the combination of (a) and (b).

Patents in this X-art collection have original classes in 75, 252, and 501. As you can see, it could be very helpful to group together the processes for making different superconductive materials. Unlike digests, X-art collections follow official patent placement procedures and may include indentation levels. These X-art collections may be found in numerical sequence at the end of each class.

Another technique used by patent examiners in the past to facilitate searching was the creation of *unofficial* or *alpha subclasses*. Here, a group of patents was selected and removed from an official subclass and then grouped into indented subclasses under the official subclass, usually with the purpose of further delineating a concept. This resulted in the original subclass being designated with an "R" for residual. Patents in an R subclass were those not specifically designated under another letter. This is best demonstrated by an example from the class 273 schedule entitled: "Amusement Devices, Games."

153 R PUZZLES

153 P ..Pyramid building

153 S ..Shifting movement

153 J ..Jumping movement

Subclass 153 R is a residual subclass and includes only those puzzles not found in 153 P, 153 S, or 153 J. Because these are unofficial subdivisions, the *Definitions* cover the entire subclass 153. As a result there are no separate definitions for these unofficial subclass divisions. Consequently a searcher cannot be sure what is included in 153 J "Jumping movement," or 153 P "Pyramid building"; therefore, all alpha designations for the same subclass should be searched for all styles of puzzle, whether searching for a jigsaw puzzle, an interlocking/unlocking key puzzle, or a Rubic Cube™ type puzzle.

Further examination of the indents under 153 illuminates the nature of these "unofficial alpha" subclasses. Search screens for this search are found in appendix 20.

153 R PUZZLES

153 P .. Pyramid building

153 S .. Shifting movement

153 J .. Jumping movement

154 . Balancing ovoids

155 . Folding and relatively movable strips and disks

156 . Take-aparts and put-togethers

157 .. Geometrical figures, pictures, and maps

158 .. Bent wire

159 .. Flexible cord or strip

160 .. Mortise blocks

161 FORTUNE-TELLING DEVICES

Subclass 153 covers all puzzles including those with rings that stack in a particular order, puzzles consisting of blocks that are joined with a mortise and tenon, puzzles consisting of wire bent into a variety of shapes, and balancing puzzles. The following examples demonstrate the breadth of patents classed to 273/153:

153 R U.S. patent 4,841,911, granted to Paul Houghton for "Variable Maze Device;"

153 S U.S. patent 4,872,682, granted to Ravi Kuchimanchi for "Cube Puzzle with Moving Faces;"

153 J U.S. patent 4,583,742, granted to Barry Slinn for "Block and Board Puzzle Game."

SUMMARY OF THE CLASSIFICATION SYSTEM

It is nearly impossible to search for patents by subject without a thorough knowledge of the classification system, including the principles used to both include and exclude subject matter from specific subclasses. To underscore this, following is a summary of the principles used to create and enlarge the classification system. This information is quoted directly from *Development and Use of Patent Classification Systems*. "Principles of the Patents Office Classification System" (*Development and Use of Patent Systems* 1966).

1. Utility as a Basis of Classification

The principal basis for classifying the useful arts in the U.S. Patent Classification System is utility, that is, the function of a process or means or the effect or product produced by such process or means. Utility as a basis of classification must be taken in the sense of direct, proximate, or necessary function, effect or product rather than remote or accidental use or application as in industries or trades. Applying proximate function, effect, or product as a basis of classification will result in collecting together similar processes or means that achieve similar results by application of similar laws.

2. Proximate Function as a Basis of Classification

Proximate function as a basis of classification is generally applied to processes or means for performing operations in which a single causative characteristic can be identified and which requires essentially a single unitary act.

3. Proximate Effect or Product as a Basis of Classification

Effect or product as a basis of classification is generally applied to complex special results of a process or means requiring successive manipulations involving plural acts.

4. Structure as a Basis of Classification

Structural features such as the configuration or physical make-up of a means may be used as a basis of classification only when the subject matter to be classified is so simple as to have no clear functional characteristics, but can only be distinguished from other subject matter by its structural features. This situation rarely arises with respect to the creation of a large group or class in the system, but frequently occurs with respect to subdivisions within a large group or class. As between a classification system based upon structure and one based upon proximate function, effect, or product, the choice is for the latter in all situations in which it can be applied.

5. Basis of Classification Applicable to Chemical Compounds and Mixtures or Compositions

A chemical compound should be classified on its structure, that is on the basis of its chemical constitution, regardless of the utility thereof. Mixtures or compositions, at least in the larger grouping, are generally collected on the basis of the disclosed utility for a particular material.

6. Analysis as a Prerequisite to System Development

The U.S. Patent System is created by analyzing the disclosures of U.S. patents and then creating classes (including the schedule of subclasses within each class) by grouping together like subject matter as represented in the disclosures of such patents.

7. Patents Grouped by Claimed Disclosure

Inasmuch as nearly every U.S. patent contains disclosure that is claimed and also disclosure that is not claimed, the general principle is that a classification system is created and a patent shall be assigned therein on the basis of that portion of the disclosure covered by the claims rather than on a portion of the disclosure that is not claimed. A disclosure that is not claimed is one that may form an element or step of a claimed combination as well as a disclosure not referred to in any claim.

8. Patents Diagnosed by Most Comprehensive Claim

The totality of a claimed disclosure must be selected, whenever possible, in creating a classification system and determining the appropriate class to which a patent is assigned, but a mere difference in the scope or breadth of claims should not make a difference in assignment.

9. Exceptions to Claimed Disclosure Principle for Assigning Patents to Specific Classes

The following situations are exceptions to the principle that a system is created and the patents assigned therein on the claimed disclosure of U.S. patents. When these exceptions are applied, it should be clearly stated in the class definitions of the classes involved.

A. Old Combination with a Specific Subcombination

Where a patent claims specific subcombination in combination with some other broadly recited subcombination, the combination and subcombination being classified in different classes, there are exceptions to the general principle that a classification system is created and a patent is assigned on the basis of the claimed disclosure; that is, the patent may be assigned to the subcombination class when all the following conditions apply:

—a relatively large number of patents are involved.

—the combination is old as a matter of common knowledge.

—no new relationship between the subcombinations is set forth.

—the other subcombination is nominally claimed.

B. Article Defined by Material from Which It Is Made

A patent for an article of manufacture, claimed by name only and in which the claim is otherwise directed to a specific material of which the article is made, is generally assigned to a class providing for the material rather than a class providing for the article.

C. Process of Utilizing a Composition

A patent claiming a process of utilizing a specifically defined composition may be assigned to the composition class where the process steps are nominally recited and the composition class provides specifically for compositions having that use.

10. Exceptions to Claimed Disclosure Principle for Patent Assignment between Subcombination Subclass and Indented Combination Subclass

Where a parent subclass has indented thereunder a combination subclass which includes as a subcombination thereof the subject matter of the parent subclass, a patent disclosing the subject matter of the parent subclass but claiming only the subject matter of the subcombination subclass is assigned to the indented combination subclass.

11. Exhaustive Division—Miscellaneous Subclass

The subdivisions or subclasses of a class in aggregate should be exhaustive, that is, they should be susceptible of receiving any future invention that may fall within the scope of the class. Exhaustive division is secured by the presence of a miscellaneous subclass.

12. Exhaustive Nature of Coordinate Subclasses: Combinations to Preceded Subcombinations

Coordinate subclasses must each be exhaustive of the classification characteristic for which the subclass title and definition provides. That is, no subsequent coordinate subclass—nor any subclass indented thereunder—should provide for the characteristic of an earlier appearing coordinate subclass. Thus, in coordinate relationship, combinations including a detail must precede subcombinations to the details, per se. A subsequent subcombination subclass receives disclosed combinations—which in their entirety are provided for in a preceding subclass—where only the subcombination is claimed; the disclosed combination is cross-referenced, if appropriate, to such preceding subclass.

13. Indentation of Subclasses

A class schedule is arranged with certain subclasses appropriately indented. In a properly indented schedule, subclasses at the extreme left in a column of subclasses are the main variants (referred to as "first line subclasses") of the class. The titles and definitions of all these first line subclasses must be read with the title and definition of the class, as if indented one space to the right under the class title.

A subclass having indented subclasses under it represents a subject divided into variants. Such subclass also includes other variants.

If no genus subclass is provided for the concepts of the several subclasses which are in fact variants of a genus, the several subclasses should be positioned in the same area of the schedule where possible, as though they were indented under the unprovided-for genus.

14. Diverse Modes of Combining Similar Parts

The classification system must recognize and provide for diverse modes of combining the same or similar parts or steps to obtain functionally (and possibly structurally) unrelated combinations.

15. Relative Position of Subclasses

The relative position of subclasses in a single class is determined by the following principles:

(1) Characteristics deemed more important for purposes of search generally should be provided for in subclasses that precede subclasses based on characteristics deemed less important. However, some subclasses of lesser importance may require precedence of position to avoid their loss from the schedule.

(2) Subclasses based upon effect or special use should precede those based upon function or general use.

(3) Subclasses which are directed to variants of a concept should either be indented under the subclass directed to such concept or precede the same, and should not form or be part of a subsequent coordinate subclass or group of subclasses.

(4) Subclasses directed to combinations of the basic subject matter of the class with means having a function or utility unnecessary for or in addition to the function or utility of the basic subject matter should precede subclasses devoted to such basic subject matter.

16. Each Class and Subclass Must Be Defined

In the U.S. Patent Classification System each class and subclass must be defined, that is the title of each class or subclass must be explained in a detailed statement setting forth the bounds of the area of subject matter for each class and subclass. A class and subclass definition must include a description of the subject matter encompassed by the class or subclass and may include any necessary explanatory and search notes.

17. Tentative Definition

A tentative or preliminary definition of a class to be created is written as soon as possible after determining the initial scope of the class. This tentative definition should be modified, if necessary, as the project progresses and as more subject matter is considered. This same principle applies to subclass definitions, that is, a tentative definition is written as soon as possible after a subclass is created and should be modified, if necessary, as more subject matter is considered.

18. Explanatory Notes for Each Class or Subclass Definition

In many instances, explanatory notes relating to excluded subject matter, the explanation of some term or expression used in the definition, statements intended to further clarify the definition, etc., may be appended either to a class or a subclass definition.

19. Search Notes for Class or Subclass Definition

To supplement or take the place of cross-referencing, search notes are needed, giving directions and suggestions for further search, setting out the relationship and lines of distinction between classes and subclasses. Search notes should indicate other classes or subclasses directed to analogous or related subject matter. Search notes should also indicate classes or subclasses directed to subject matter constituting either a combination or subcombination of the class or subclass in which the note is written. However, care should be taken when writing a search note indicating a class directed to a subcombination which is common to subject matter of several classes. The Index to Classification is a useful guide in locating such subject matter.

20. Cross-Referencing

Nearly every patent discloses subject matter that is classifiable in a different class or subclass than that which provides for the subject matter of the controlling claim. In the U.S. Patent Classification System such different subject matter is appropriately provided for by the assignment of one or more cross-references. Such subject matter falls into two categories, (1) subject matter which is separately claimed, per se, in a claim other than the controlling claim and (2) subject matter which is disclosed but not claimed, per se, in a claim other than the controlling claim.

21. Cross-Referencing Claimed Disclosure

Where a patent has several claims which if separately found in different patents would effect assignment of such patent in different subclasses, either in the same or in different classes, original assignment of the patent is on the basis of the most comprehensive claim as between classes and schedule superiority within a single class. It is obligatory in such instances to cross-reference the patent to the subclass or subclasses providing for the subject matter of such other claims, unless search notes are provided which would lead a searcher to the subclass to which the patent is assigned on the basis of the most comprehensive claim.

22. Cross-Referencing Unclaimed Disclosure

Any disclosure in a patent which is disclosed but not claimed, per se, may be cross-referenced into any part of the classification system at the discretion of the Classifier. The following criteria should be considered for such cross-referencing: (1) the disclosure must, in the best judgment of the Classifier, be novel and (2) the disclosure must be of sufficient detail and clarity to be useful as a reference. No cross-reference is made when a search note is appended to the definition of the subclass eligible to receive the cross-reference, indicating that the subclass containing the original must be searched.

23. Superiority Among Classes

A. With respect to an application or patent directed to one claimed disclosure, assignment is to the class that is the locus of the prior art for the same subject matter. The identity of the proper class is established through study of class definitions and notes of classes suggested by the Index to Classification or lists of classes or by personal knowledge of the location of the prior art.

B. With respect to an application or patent including claimed disclosure to diverse inventions, the principles listed below must be considered and applied, if appropriate, step-wise, in the order listed, to select the single disclosure that will control assignment as in "A" above.

(1) Most comprehensive claimed disclosure governs.

(2) Order of superiority of statutory categories of subject matter.

(a) Process (of using product b, for example using a fuel or radio transmitter).

(b) Product (of manufacture, e.g., a fuel or radio transmitter).

(c) Process (of making product b, for example fuel).

(d) Apparatus (to perform c or to make b, for example machine, tool, etc.).

(e) Material (used in c to make b, for example an alloy, machine tool).

(3) When, and only when, principles 1 and 2, given above, fail to solve the question of the controlling class, the relative superiority of types of subject matter as shown by the following list is used.

(a) Subject matter relating to maintenance or preservation of life is superior to subject matter itemized in b-d below.

(b) Chemical subject matter is superior to electrical or mechanical subject matter.

(c) Electrical subject matter is superior to mechanical subject matter.

(d) Dynamic subject matter (i.e., relating to moving things or combination of relatively movable parts) is superior to static subject matter (i.e., stationary things or parts nonmovably related).

24. Superiority Within a Class

Where different subclasses of the same class are involved, the patent will be assigned to that one of several subclasses defined to receive the several claimed inventions which stands highest to the schedule of subclasses.

CONCLUSION

The purpose of the classification system is to reduce the number of individual search problems by associating related art units (subject matter) into large groups, subdivide the large groups into subgroups, and arrange the smaller subgroups in a sequence or pattern. In this system each large group is called a class, each subgroup a subclass, and the sequence or pattern arrangement of the subclasses is a schedule. The classification system, which includes the class definitions and schedules, serves to organize patentable ideas and specific information found in patents. In doing so, it provides for storage and retrieval of the prior art (subject matter patentable under U.S. patent law) so that it can be searched by historians, librarians, inventors, and others for information stored in these patents and used to search for negative information—ideas not yet patented.

The importance of this system cannot be overstated; information from this chapter will be put to use in later chapters describing the search process. A clear understanding of this system, its rules, strengths, and oddities, will make any patent subject search easier, less frustrating, and more accurate.

The next chapter begins the search process by defining the different types of patent searches.

REFERENCE

Development and Use of Patent Classification Systems. 1966. Washington, D.C.: U.S. Patent and Trademark Office, 3-8.

5 Defining the Search

To philosophize is to generalize, but to generalize is to omit.
—Oliver Wendell Holmes

Before beginning any patent or patent related search it is a good idea to decide what type of search will produce the best results for the particular problem at hand. This chapter consists of short definitions of the types of searches that are possible. As discussed in the introduction to this book, there are many reasons for doing a patent search; however, most can be categorized as either legal or information searches. Legal searches are concerned with discovering and evaluating the protection granted by the patent, while information searches are concerned with whether specific information can be found in a granted patent. Both types of search are defined in this chapter because it is useful to be aware of all uses of the patent file, even though the focus of this book is on information and patentability searches.

All of the defined searches may be carried out in the Public Search Room at the Patent and Trademark Office in Arlington, Virginia, and many aspects of bibliographic, state-of-the-art, and subject searches can be carried out in any patent depository library (PDL) listed in appendix 13. Each PDL's collection can vary quite a bit; some have complete collections of patents, some have less extensive collections, and a few others even have a collection of foreign patents. It is a good idea to check with specific PDLs about the size and nature of their collections.

It should be noted that while it is the intent of this book to guide the potential searcher through the process, it is possible to hire professionals to conduct the search. Professional searchers generally charge by the hour; lists of qualified attorneys and agents are available in all PDLs.

Even if you decide to hire a professional, an idea of the complexity and nature of the process can help you communicate with that professional and save money. Some of the questions to ask are: How many patents has the professional searcher successfully applied for? What area are these patents in? What type of technical background does the searcher have? Ask to look at a number of completed applications. Obviously not all inventors or searchers should undertake the task of applying for a patent without some type of help.

TYPES OF SEARCHES

Bibliographic Searches. This type of search is the easiest and the quickest because the searcher already has a patent number or an inventor's name. The point of this type of search is to find out what was covered by a specific patent number or to find out what patents a particular inventor has to his or her credit. Bibliographic searches can be done as a part of historical, biographical, archaeological, or product research. This type of search is discussed and illustrated with sample searches in chapter 7.

Patentability Searches. Patentability searches are probably the most common type and are covered in detail in chapter 8. A patentability search attempts to determine whether a specific invention is (1) within the scope of patentable subject matter, (2) useful, (3) new or novel, and (4) unobvious.

While patentability searches can be performed early in the development of an invention, they are more commonly done prior to submitting a patent application. The purpose of this type of search is to determine whether there are any previous patents (prior art) that might prevent the searcher from patenting his or her idea. Another benefit is that the searcher can be spared the expense of filing an application, since the filing fee is not refunded if the application is rejected. A search may also turn up prior art that might be useful in preparing the application.

State-of-the-Art Searches. Any search designed to give an overview of prior art (technology) in a specific area is called a state-of-the-art search. This is basically an information gathering approach and can be as extensive or cursory as needed. This type of search is done to solve a specific problem, find technology to license, determine what the competition is up to, and for other similar purposes. For example, a manufacturer might conduct a state-of-the-art search to determine what has previously been done, to determine if an area has been overlooked, or to determine if it has had a lot of patenting activity. This type of search is very common.

Continuing Searches. A continuing search is nothing more than a current awareness search of recently issued patents. Generally these searches are of two types: keeping up with patent activity in areas of interest and keeping up with competitor's activities. This can be done online (see chapter 10) or by scanning the *Official Gazette* weekly under subclasses of interest. It is very common for companies to scan the *OG* in just this way.

Assignment Searches. When a patent is assigned to another person or company it is the same as a sale, even though the terminology is just a bit different. For example, the buyer, whether a company or individual, is called the "assignee" and the seller (often the inventor) is called the "assignor."

Assignment records of issued patents are public. However, licenses are not recorded in the Patent and Trademark Office. The purpose of this type of search is to determine legal ownership of a patent.

Infringement Searches. An infringement search is used to determine if proposed activity might infringe on or be covered by unexpired patents. This type of search is concerned only with the claims of unexpired patents. As a legal issue that can be adjudicated in the courts, infringement would be pursued by a patent attorney. Only the Patent and Trademark Office in Washington has the complete record of all office actions as they relate to an individual patent. This complete record is referred to as the file wrapper; it contains copies of correspondence between the office and the inventor, information on classification developed during examination, index of the claims, field of search, and search notes. These search notes can be extremely useful since they can contain references not printed on any resulting patent, including notes of any consultations with other examiners and results of library and literature searches. This information completes the application file and records areas or documents considered by the examiner. For these reasons infringement searches are rarely done in a patent depository library.

Validity Searches. Validity searches are generally undertaken by companies or individuals trying to determine if it is possible to invalidate another's patent. The searcher is looking for issued patents that anticipate or make obvious another issued patent. The searcher could also be looking for earlier public knowledge on use of the invention, technical errors, fraud, or anything that would cause the patent to be declared invalid. As a result, a validity search is done to determine if an unexpired patent is valid and enforceable. This type of searching is generally done by patent attorneys.

Right-to-Make Searches. A right-to-make search concentrates on expired patents, unlike an infringement search, which concentrates on unexpired patents. These searches are done to determine if another company's process, product, or design has expired and can then be copied with impunity.

CONCLUSION

Bibliographic, state-of-the-art, patentability, right-to-make, and continuing searches can be done in PDLs by knowledgeable searchers; however, assignment, infringement, and validity searches generally are within the purview of attorneys and as such are infrequently done in a PDL.

The following chapters explain and describe the basic steps and tools used to perform bibliographic, continuing search, patentability, and state-of-the-art patent searches. Each of these types of search is described in detail and demonstrated through the use of worked examples.

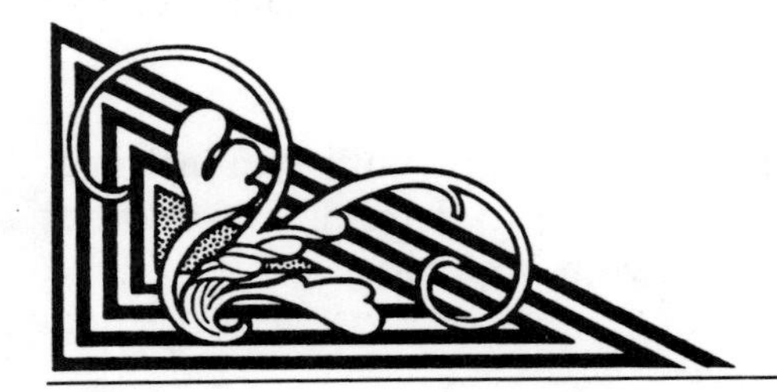

6 The Search Begins

Though this be madness, yet there is method in't.
—Shakespeare

This chapter consists of an overview of searching by bibliographic data: U.S. patent number, foreign patent number equivalents or patent families, and inventor's name or assignee's name. These are some of the easiest searches to do because the searcher has very specific information. They can be done manually (except for patent family) or on any of several commercial patent databases. See chapter 10 for more information on searching these databases.

U.S. PATENT NUMBER

References to specific patent numbers can be found in a variety of ways and places. The reference might result from a literature search, a footnote, a bibliography, or even an advertisement or product catalog. The number might also be found on the product itself, or be a cross-reference found in a previous subject search.

Access to bibliographic information by patent number is possible through:

1. The *Official Gazette (OG)*, which is owned by many public, college and university libraries. See appendix 21 for examples of the *OG* and chapter 14 for a short description of the *OG*.

2. Patent databases such as CLAIMS, LEXPAT, and WORLD PATENT INDEX (WPI). These are discussed in more detail in chapter 10.

3. Copies of U.S. patents are available at the Public Search Room of the U.S. Patent and Trademark Office, at patent depository libraries, or from commercial vendors.

4. Copies of foreign patents by number can be obtained from the relevant national patent office, commercial vendors, or from the British Science Lending Library in London.

SAMPLE SEARCH: PATENT NUMBER

The July 1988 issue of *Atlantic Monthly* stated that U.S. patent 3,595,339 was expiring. No indication was given as to the inventor's name or the title of the patent. All that was said was "A pair of stilt boots, incrementally vertically adjustable, whereby the operator [can] raise or lower them by selective weight shift and foot pressure control" (*Atlantic Monthly* 1988).

If you wanted to know the inventor's name, the patent's title, and if it was assigned to anyone at the time of issuance, you would take the following steps:

Step 1A: Go to the *OG* and look up the patent number. Generally the range of patents covered by each volume is printed on the spine.

a. The patent number appears in the July 27, 1971 issue.

b. The inventors were Wesley D. Ballard and John H. Staley, both of Waco, Texas.

c. It was not assigned to anyone at the time of issuance.

or

Step 1B: Check commercial online databases or CASSIS (online or CD-ROM) for coverage and enter the patent number. This number could have been entered in WPI and CLAIMS databases.

Step 2: Examine complete patent if truly interested.

The complete patent would describe how this invention worked; provide drawings of the invention; give the class, subclass, and assignee listed at time of issuance; and describe the legal protection given this invention. It would also list any related patents known at the time the patent was granted.

FOREIGN PATENT NUMBER

References to foreign patent numbers can be found in the same way as references to U.S. patent numbers. The following are examples of foreign patent citations:

"Water-thinned, air-drying aerosol paint compositions." Guenther Pezina and Andreas Mandel. Ger. DE 3,808,405.

"Testing water." Valerie Anne Argent. Brit. UK Pat. Appl. GB 2,215,458.

"Silver halide color photographic material containing hydrazine derivative." Keiji Obayashi and Morio Obayashi. Jpn. Kokai Tokyo Koho JP 01,147,455.

Unfortunately most public and university libraries do not have access to foreign gazettes or foreign patents. Consequently, it is sometimes necessary to determine if a particular foreign patent has a U.S. equivalent.

The best search tools for determining equivalencies/families are such online databases as WPI or INPADOC's Patent Family Service. An overview of foreign patents can be found in chapter 11, "Foreign Patents" and equivalency/family searching with an example is covered in detail in chapter 10.

ASSIGNEE'S NAME

An inventor or assignee search is done if the searcher wants to see what a particular person has invented or what patents were owned at the time of issuance by a particular company. Both manual and online searching by inventor/assignee are available. It is important to remember that it is generally not possible to search inventor's names online with any degree of confidence prior to 1975. Check database literature for information on coverage.

Searching by company names can be problematic. Some companies such as IBM take out a large number of patents every year; consequently, it can be time consuming to examine every patent issued. Even more discouraging is that the Patent and Trademark Office makes no attempt to control or standardize company names. As a result, the company name on the patent face may be derivative or that of a subsidiary. Another problem is that patents are not always assigned at issuance to the company expected. Directories, such as Dunn's *Million Dollar Directory, Thomas Register of American Manufacturers, Directory of Texas Manufacturers, U.S. Industrial Directory, Directory of Corporate Affiliations,* and *Who Owns Whom* can be useful in determining subsidiary names or associated companies.

Searching by a particular assignee's name can be accomplished in several ways. The first step is to determine, if possible, the probable years of interest, and then consult one or more of the following sources:

1. The *Official Gazette*: The *OG* lists inventors of patents issued each week, but a search through the weekly issues can be very time consuming.

2. The *Index of Patents Issued from the United States Patent and Trademark Office*: Part I, "List of Patentees," is an annual alphabetical compilation of patentees (inventors) and assignees (owners) at the time of issuance. (See appendix 22.) Part I has been published since 1926. Earlier indexes are included in the *Annual Report of the Commissioner of Patents* 1790-1925. The *Index* is published quite slowly; 1984 was not available until early 1987.

3. "Patentee/Assignee Index" (U.S. Patent and Trademark Office): This microfiche publication lists in alphabetical order five to six years' worth of patentees and assignees at the time of issuance. A quick search to determine company patenting activity is possible. The "Index" is updated quarterly.

4. Assignee searches for differeing time periods are also available using commercial databases or CASSIS CD-ROM. See table 10.1 on page 77 for dates covered.

SAMPLE SEARCHES: ASSIGNEE

In the early 1980s Texas A&M University was assigned a patent that dealt with a fault detection device. The problem is to find this patent.

Step 1: Go to the microfiche *Patentee/Assignee Index* for 1978-1985. (If this is not available check the *List of Patentees* for each year from 1980 until you find the patent.)

Step 2: Find the microfiche card that contains entries for Texas A&M. Follow the same procedure if using the annual cumulations of the *List of Patentees*.

Step 3: Find the following names under both Texas A&M University and Texas A&M University System:

1980, no entries

1981, Tsutsui, Tomo; Durso, William S.; and Kajo, Saksi

1982, Lawthon, James T.; Milberger, Lionel; Denk, Eddie W.

1983, no entries

1984, Chiou, George; Dhar, Hari P.; Bockris, John O.; Russell, B. Don Jr.

Step 4: Examine the titles and discover that the Russell patent has the title "High Impedance Fault Detection Apparatus and Method," and the number 4,466,071.

Step 5: Check the *Official Gazette* entry for confirmation and decide if this is it.

This search was quite easy and required little time. The following search is a better example of what happens all too frequently.

A searcher wants a copy of a patent for a washing machine relay switch invented in the late 1970s by General Electric.

Step 1: Go to the microfiche Patentee/Assignee Index for 1978-1985.

Step 2: Look up General Electric and pull the three applicable microfiche cards.

Step 3: Rethink the search; there are too many entries to review. In 1987, for example, General Electric was granted 779 patents.

Step 4: Searcher decides that search is too difficult and time consuming and gives up. The searcher was not interested enough to try a subject search on relays, which would have eventually answered the question.

Many searches will end in just this way—the amount of work necessary to complete the task is just not worth the effort. It should be noted again, that any assignee search in a PDL will only retrieve the assignee (owner) at the time the patent was issued.

Another way to do assignee searches involves using one of the commecial databases such as WPI, LEXPAT, CLAIMS, or CASSIS CD-ROM. This type of searching is discussed in more detail in chapter 10. However, online assignee search is very similar to corporate source searching. Depending on the database selected, the searcher can search for assignees by year and cross this with inventor's name if known. Just like any manual searching for assignee, the online databases list only the assignee printed on the face of the patent when it was issued.

The best way to become familiar with conducting patent number or assignee searches is to do one. Please try the following searches, using the steps described and demonstrated in this chapter. The answers, with discussion, are found in chapter 13, "Answers to Problems."

PROBLEMS

1. The number D 281,087 was found on the underside of a small plastic car. Is this a patent number, for what country, and if it is a patent, what does it protect?

2. An archaeology student doing field work at a state historical site has discovered a small metal box with a funnel shaped opening. On the bottom is the number 1,090,018 and "Tampa." Is this a U.S. patent number, and if so, what is it?

3. What country begins its patents with the prefix SU?

4. A recent television commercial for Mercedes-Benz mentions that the company has had a patent since 1951 for a safer body frame. In this advertisement Mercedes stresses that because this invention was so important it had never prosecuted anyone for patent infringement on the invention. The problem is to discover if Mercedes-Benz received a U.S. patent in this time period for a car body frame.

CONCLUSION

The bibliographic data searches covered in this chapter are quite common. Citations to patent numbers, assignees, or company names can be found in many places such as magazines, newspapers, and books. This type of searching can be done manually or online in many libraries and generally requires no special tools* or knowledge of the classification system.

The next chapter discusses an even more common type of bibliographic search, the search for patents by a known inventor. While this search is similar to an assignee search, it does have a number of specific problems. Rarely does the citation give a complete name, title, or date of issuance. These and other problems, along with a sample search, are covered in chapter 7, "Sample Search for a Known Inventor."

REFERENCE

"Monthly Notes." 1988. *Atlantic Monthly* 253 (July): 10.

---

*A complete list of the tools needed to do a patent search is found in chapter 14.

7 Sample Search for a Known Inventor

Certainly an inventor ought to be allowed a right to the benefit of his invention for some certain time. Nobody wishes more than I do that ingenuity should receive liberal encouragement.

—Thomas Jefferson

Chapter 6 covered a number of typical bibliographic patent searches, including searches by patent number, assignee, and patentee or inventor. But by far the most common bibliographic search is by inventor's name. A magician might wish to know if the magician and escape artist Harry Houdini had ever patented any of the devices used in his act, another searcher might wish to find the patents of George Washington Carver, a researcher might wish to discover if presidential scholar and mechanical engineering professor Ilene Busch-Vishniac has any patents to her credit, and an architectural historian might wish to know if Frank Lloyd Wright ever had a patent.* This chapter includes an explanation of how to conduct a search by the inventor's name along with a sample search demonstrating the techniques used to search by an inventor's name.

THE SAMPLE SEARCH

References to inventors can be found in a number of places other than scientific or technical papers. Sometimes the most difficult patents to find are those cited in popular magazines and newspapers. These references rarely contain an exact date, title, or patent number; often only the inventor's name and some idea of the subject are included. Our sample search is a good example of a reference without much information. The February 1988 issue of *Texas Monthly*, contained the following statement: "... making for a flamboyant clan that has given contemporary Texas such necessities as the Styrofoam coffee cup, [which was] invented in 1956 by Jimmy Harrison" (*Texas Monthly* 1988). Our job is to find out if Jimmy Harrison really invented styrofoam coffee cups, or anything else.

*Patent 1,370,376, "Diver's Suit" (permitting escape), was granted to Harry Houdini in 1921. Many patents were granted to George Washington Carver, including 1,522,176, granted in 1925 for "Cosmetic and Process of Producing the Same." Ilene Busch-Vishniac has also been granted a number of patents, including 4,558,184, granted in 1986 for "Electret Transducer for Blood Pressure Measurement." Frank L. Wright had a series of glass block designs (D 27,978 to 28,017) patented in 1897 and assigned to the Luxfer Prism Patents Company in Chicago.

Step 1: Evaluate the clues:

a. Determine patentee's name. Remember that the Patent and Trademark Office requires full name including at least one initial.

b. What does the date mean?

Did he think of the idea in 1956?

Did he apply for a patent in 1956?

Did he receive a patent in 1956?

Based on our evaluation of the clues, we made the following assumptions:

a. He applied for and received a patent around 1956.

b. His name is James something Harrison.

c. His name might also be Jimmy something Harrison.

d. He was living in Texas at the time his patent was granted. This is important. While it is a well-known fact that once a Texan always a Texan, he could be a misplaced Texan. The Patent and Trademark Office is not interested in geographic loyalty. For the purposes of a U.S. patent, the inventor's address is his residence when the patent was issued, not what he considers to be his home.

Step 2: Begin the search:

a. Go to the 1956 *Index of Patents*.

Look up his name in both the utility and design patentee sections of the *Index*. This tool has two separate name indexes, one for utility patents and one for design patents. It is possible that his patent for the styrofoam cup could be considered a design.

Find patent 2,734,518, James A. Harrison, "Machine for Cleaning Pipettes and Petri Dishes."

Find patent 2,764,860, James M. Harrison, "Plug for Test Tubes and the Like."

b. Repeat process in the 1957 and 1958 *Indexes* since the ones found in 1956 do not seem relevant.

Find in 1958 patent 2,828,509, James M. Harrison, "Plastic Molding Machines." This is a possibility, so move on to Step 3.

Step 3: Check 2,828,509 in the *Official Gazette*.

a. Check for a Texas address.

b. Read the printed claim; it may help determine if this is the Harrison patent we are looking for.

Since nothing so far specifically excludes this patent, move on to Step 4.

Step 4: Check the complete patent in a patent depository library or by ordering a copy from the Patent and Trademark Office or a commercial supplier such as RAPIDPAT or Airmail Patent Service, both of which are in Arlington, Virginia. Once the patent is in hand, do the following:

a. Read the disclosure.

b. Read all the claims.

c. Check for prior art citations.

Step 5: Make a decision as to whether or not this is the patent we are seeking.

Is this the patent for styrofoam cups? The patent we found is the only one that comes close to having all of the elements mentioned in our citation. The date is about right, the address is Ft. Worth, and the subject is still a possibility. But on reflection, another possibility suggests itself. Is it more likely that James/Jimmy Harrison invented a process for creating styrofoam cups rather than the cup itself? One determining factor might be to see if or when the word "styrofoam" became a trademark. This can be done by checking the annual index of the U.S. Patent and Trademark Office *Official Trademark Gazette*, the *Trademark Register*, or the online file Trademarkscan™. A Trademarkscan search shows that Styrofoam™ was first used as a tradename on September 11, 1945, by the Dow Chemical Company of Midland, Michigan.

Returning to the original question—did a Texan invent styrofoam coffee cups—how confident are we in our search? Our research shows that it is highly likely that Jimmy Harrison invented the process that made possible styrofoam cups. A copy of the *OG* entry for this patent is found in appendix 23. This search was done manually because the primary databases (Derwent, WPI, and INPADOC) do not cover the 1950s.

Another example of a search by an inventor's name is more historical. A gentleman comes into the library and says he wishes to verify an old family story. It seems that his uncle, Leo Chapman, had told the family for years about his invention of a way of making printed ledger paper. Part of the story was that he held a number of patents, but this was the only "good one." Our gentleman wants to know if his uncle actually had a patent. He knows the time period would be sometime in the 1920s and that his uncle was a linotype operator living in Chicago.

Step 1: Go to the *Index of Patents* and look up the name Leo something Chapman.

a. Find an entry for a Leo M. Chapman in 1925 for U.S. patent 1,539,935, "Hairpin."

b. Continue searching in the *Index* for 1926, 1927, and 1928.

c. In 1928 find cross-reference from William C. Hollister to Leo M. Chapman for patent 1,685,514, "Means and Method for Producing Vertically-Ruled Printing Forms."

Step 2: Go to *Official Gazette* for 1928 and examine patent 1,685,514 for city and assignee. After examining the *OG*, our patron remembers that his uncle worked for Lino-Tabler Company.

Step 3: Patron decides that the family story is essentially correct. His uncle did invent a type of ledger paper. A copy of the *OG* entry for this patent is found in appendix 24.

The final example demonstrates how even a seemingly simple inventor search can turn out to be complex and that many patent searches can also utilize traditional library tools during the course of a search.

A patron remembers reading in a newspaper that King Hassan II of Morocco was the first king to receive a U.S. patent. She believes that the invention involved a heart monitor and that it was granted sometime in early 1989. The search could have been done looking at each weekly issue of the *OG*, but because the question was asked in early 1990 this would have been extremely time consuming. Therefore an online search was selected.

Step 1: Select DIALOG File 25, January 1982-October 1989.

Step 2: Input author's name (see figure 7.1).

Step 3: Examine each Hassan in the *OG*. The results are negative. None of the listed Hassans had a patent on anything remotely related to the heart and none were the King of Morocco.

Fig. 7.1. Search for a known inventor on DIALOG File 25, January 1982-October 1989. Reprinted with permission of Dialog Information Services, Inc.

e au = hassan

Ref	Items	Index-term
E1	1	AU = HASSAL STEPHEN J
E2	2	AU = HASSALL THOMAS JR
E3	0	*AU = HASSAN
E4	1	AU = HASSAN AHMED A
E5	1	AU = HASSAN AWATIF E
E6	1	AU = HASSAN JAVANTHU K
E7	1	AU = HASSAN KAMAL-ELDIN
E8	1	AU = HASSAN MELVIN
E9	3	AU = HASSAN MOHAMED A
E10	3	AU = HASSAN MORRIS
E11	1	AU = HASSAN SHAWKY A
E12	1	AU = HASSAN WORTHY O
E13	1	AU = HASSANALI-WALJI AHMED

Step 4: Return to CLAIMS (File 25) and input Hassan's name as if King were his last name. Results are negative again.

Step 5: Read section 605 of the *MPEP*, "Applicant." This section clearly states that "a patent must be applied for in the name of the actual inventor or inventors. Full names must be stated, including the family name and at least one given name without abbreviation together with any other given name or initial" (*MPEP* 1989). Therefore:

Step 6: Look up King Hassan in an encyclopedia and almanac to determine if he has a family name. Because the results are negative, call a major academic library reference desk. If they cannot discover a family name, go to Step 7.

Step 7: Since the patron thought she read this in a newspaper, do an online search of newspapers using his name and the word patent. Find that the *New York Times*, Saturday, February 25, 1989 states that "King Hassan received patent 4,805,631" (*Heart Study* 1989).

Step 8: Check 4,805,631 in the *OG* to see what name was used, and you will discover that this number corresponds to "Device for the Detection, Study and the Supervision of Diseases, and in Particular Heart Diseases, Resulting in Electrically Recordable Manifestations." It was granted to Sa Majeste H. Roi Du Maroc, II. Go back to File 25 of CLAIMS and input the patent number to see how the inventor's name was listed. It was listed as Du Maroc Sa Majeste Roi II.

When this search actually was done in a patent depository library, the library staff called the Patent and Trademark Office and asked why the name appeared this way. The file wrapper on this patent was read and it was discovered that this is how King Hassan signed his inventor's oath. Due to our question a correction is expected soon. This correction will be listed in the *Official Gazette*.

These three searches by inventor name are common types. The first demonstrates a search to find out more information based on a citation to an inventor, the second demonstrates a historical search, and the last demonstrates the use of other tools, along with patent tools, to get an answer.

The best way to become familiar with anything is through experience. Please try the following inventor search on your own by following the same steps and using the same tools as those in the Jimmy Harrison and Leo Chapman examples. While the answers are not as important as the technique for finding the answers, the solutions to these questions can be found in chapter 13.

PROBLEMS

1. A recent book by Kenneth A. Brown, *Inventors at Work* (Tempus Books, 1988) states that Jerome Lemelson has more than 400 patents and "is one of the most productive inventors in the United States." Determine if he has "a lot" and find several examples.

2. Buckminster Fuller is credited with inventing the geodesic dome. Did he get a patent on it?

3. Does rock star Eddie Van Halen have a patent?

Try to place the inventor in a time period because this will give you a starting point for using the tools. The tools to use are the *Index of Patents, Patentee/Assignee Index*, or one of the commercial databases such as INPADOC, WPI (World Patent Index), or CLAIMS.

CONCLUSION

This chapter discussed and demonstrated inventor searching. The techniques covered are used to find patents when the searcher has an inventor's name. These same techniques could be used to find patents by famous people such as Albert Einstein, Abraham Lincoln, and many others. They can be used to document the creativity of individuals, find additional information such as that in the Styrofoam cup example, or verify family stories, as in the Leo Chapman example.

The next chapter covers searching for patents by subject and is what most people mean when they talk about doing a patent search. The next chapter begins to put together information covered in previous chapters, for example, patents as documents, the classification system, and how to use the patent tools. These are used to perform subject-of-invention or patentability searches.

REFERENCES

"Heart Study System by King Hassan." 1989. *New York Times*. Sec. A, col. 5 (February 25): 36.

Jubera, D. 1988. "The Promised Land." *Texas Monthly* 16 (February): 166-69.

Manual of Patent Examining Procedure (MPEP). 1989. "Applicant." Washington, D.C.: U.S. Patent and Trademark Office.

8 Subject-of-Invention Searching

The issue of patents for new discoveries has given a spring to invention beyond my conception.

—Thomas Jefferson

What most people refer to as a patent search is in fact a subject search; other types of searches such as inventor, assignee name, or patent number involve finding information from an access point other than subject. There are two basic reasons for performing a subject search. The first is to determine if an idea has been previously patented. This type of search is sometimes referred to as a patentability, novelty, or pre-ex search. The second reason, often referred to as a technical or state-of-the-art search, is to determine if a solution to a problem has been discovered, to find a new use for an old product, or to find information that would help in developing new products. The only difference between these two searches is in the desired outcome. In the first case, the inventor does not want to find anything remotely related to his or her invention, and in the second case the searcher hopes to find existing information or solutions to the problem. In actual fact, both of these searches are conducted in the same way. The major difference is in the degree of specificity possible in describing the precise nature of the invention. When doing a patentability search it can be assumed that the inventor is familiar with how his or her invention works and what it does. Consequently, the inventor should be able to describe it in fairly concrete terms and should recognize any existing patent that is similar. However, a technical search can be more difficult to formulate precisely, since the searcher may not be able to describe exactly the solution another inventor has used to solve the searcher's specific problem. In either case, the same steps are used.

This chapter consists first of an outline of the steps used in any subject search and then of a discussion of the tools and how they are to be used.

OUTLINE OF SUBJECT SEARCH STEPS

I. *Knowing the Idea or Invention*
   A. List all essential parts or steps.
      1. Function
      2. How it works
      3. How it is structured
      4. Results produced
      5. Steps involved in the process
   B. Determine the scope of the invention (what was really invented: a device, a process, an improvement, etc.).

C. Go to the following tools in the order listed.
  1. *Index to U.S. Patent Classification:* an alphabetical listing of subject terms, phrases, trademarks, synonyms, and acronyms
  2. *Manual of Classification:* the schedules for all classes in numerical order
     a. Subject matter is *hierarchically* arranged
     b. The order in which the subclasses appear in a class establishes the order of superiority between concepts
     c. The number assigned to a subclass title has no function other than to act as an "address"
     d. Location within the schedule *hierarchy* is paramount
     e. Review main line (all caps) subclasses
     f. Review indented subclasses (each dot represents another level of indentation)

II. *Verifying or Modifying a Search (field of search)*
  A. Go to *Classification Definitions*.
     1. Read entire "Statement of Class Subject Matter"
     2. Review "Head Notes"
     3. Review "Cross Notes"
     4. Go to subclass definition
        a. Review notes
        b. Review "Search This Class, Subclass" notes
        c. Review "Search (other) Class" notes
        d. Review search notes in outdents (those comments appropriate to all indented subclasses)
     5. Get list of patents in field of search (class/subclass)
        a. CASSIS online or CD-ROM (available in PDLs) is updated quarterly and contains additions and changes in classification since the last update.
        b. *Official Gazette* lists patents issued in each class/subclass week-by-week, and is then cumulated into the *Index of Inventions*. However, it should be noted that class/subclass designations are not static and can be changed at any time.
        c. Other online retrieval services such as INPADOC, WPI, and CLAIMS.
  B. Check selected patent in the *Official Gazette*; the summaries can be used as a quick check that you are on the right track. The point of this step is to determine if the patents stored in your selected class/subclass combinations are what you expected to find. If they are not, go back to II and try to find combinations that better describe your subject. If you are on the right track, continue.
  C. Check the complete patents.
     1. Look at OR patents first, if not pertinent:
     2. Look at XR patents
        a. Review "Head Notes" and "Cross Notes"; "Subclass Definitions" and "Search Notes"
        b. Expand or narrow field of search accordingly
     3. If you do not think you have found the right location but an XR patent number seems very relevant (i.e., you have found one or more pertinent patents):
        a. Determine their classification and repeat II.
        b. Review the patent's field of search (on first page of any patent)
        c. Review cited patents and determine their classifications (on first page and in body of the disclosure)

III. Iterative Process (continuous modification of search as more information becomes available)

A. If after several passes you have found no patents remotely similar to your invention, try:

1. A keyword search on CASSIS (online or CD-ROM). Remember that only the last two or three years are available (no manual keyword search of patent abstracts or titles is available).
2. Keyword search on a commercial patent database such as CLAIMS, LEXPAT, WPI, or PATSEARCH.

SEARCH TOOLS

As we go through the specifics of using the following search tools, please keep in mind the generalized outline just covered. The intent of this section is to describe the tools and how they are used and then demonstrate this use through examples.

After analyzing the elements of your invention or determining what type of information is desired, it is time to use the *Index to the U.S. Patent Classification* (see appendix 25).

The *Index* consists of an alphabetical arrangement of common, informal, natural-language subject headings or terms. Use it to look up those terms that best describe the topic. If you do not locate anything at first, look for terms with approximately the same meaning, for terms either broader or narrower in scope, or for terms that represent a different approach to the subject.

When, for example, you look up *pencil* in the *Index* you will find something like this:

	Class	Subclass
Pencil	401	49+
Clip	D19	56
Compositions	106	19
Sharpeners	30	451+

The class/subclass numbers given are suggested places to begin searching the *Manual of Classification*. Subclasses with a plus (+) sign, such as 49 in the pencil example, are intended as a reminder that this is only a starting point. The searcher should continue scanning down the schedule until a relevant subclass is found.

Other important designations may be found in the *Index*. The *D* in class 19 is a reference to a design classification. In this example, ornamental pencil clip patents will be found in class D19. The *D* designation is a signal to look in the design patent schedules; these are filed in a separate section from the other (utility) schedules. Our example does not illustrate another possible, but relatively rare, type of marking, the asterisk (*). The * designates a collection of related patents that are not classed together. For example, class 493 subclass 945* was found in the *Index* under "Coin and Coin Handling, Wrapper or holder"; 945* should consist of patents from a number of various classes and subclasses. The * sign denotes a cross-referenced art or X-art collection.

Another possible delineator is found in the next example, the alpha delineator (another alpha example is discussed on p. 28). The alpha delineator notifies the searcher that this subclass has been divided. In the next example, subclass 26 has an *R*. This means that subject matter in subclass 26 has been divided and the R or residual subclass contains patents on hammer mills.

When using the *Index*, the idea is to look up many different terms until you find the ones that seem most appropriate. For example, in the *Index* you will find a reference given for the tradename "frisbee." When you follow the frisbee™ reference to the *Manual of Classification*, you will see that frisbee becomes an "aerodynamically supported or retarded spinning disc." Since the *Index* is only used as a starting place for the more technically descriptive *Manual*, you should not become too concerned with finding exact terms. The idea is to get close, since the *Manual* and subsequent tools will provide more exact definitions and descriptions of the invention being sought. When using the *Index*, it is important to remember that words can have different usages. The next example from the *Index* demonstrates the different usages of "hammer" and also why the user may need to look up a number of different terms that are either synonymous or highly related to the ideas being searched.

	Class	Subclass
Hammer	81	20+
Burglar Alarm	116	88+
Claw	254	26R
Mills	241	185R
Musical Instruments		
Piano	84	236+
Stringed Instrument	84	323+

This example from the *Index* shows many of the possible places hammer and hammer-like inventions could be classified. Please notice that "Musical Instruments" is indented under "Hammer"; consequently these hammers are used as part of a musical instrument and those used specifically in pianos are found in 84/236 and beyond. However, if looking for the tool "claw hammer," you would start in class 245 subclass 26R.

The *Index* is published every December, while the *Manual of Classification* is updated regularly as needed. In other words, the *Manual* is updated as the technology being classified changes, not on any set schedule. As a result, it is possible to find a reference in the *Index* to a class or subclass that no longer exists. A number of strategies can be used to overcome this difficulty, including online searching, examining change orders, and scanning relevant classes in the *Manual*.

USING THE *MANUAL OF CLASSIFICATION*

The *Manual* comprises numerically ordered lists known as schedules for all classes and subclasses. See appendix 26 for examples of each type of schedule. To continue the search on pencils in the *Manual*, first note down the reference obtained from the *Index*, in this case to class 401, subclass 49. Looking up 401/49 in the *Manual*, you will find the title of this class is as follows:

401 COATING IMPLEMENT WITH MATERIAL SUPPLY

49 .Solid material for rubbing contact

The language used in the *Manual* is more technical and descriptive than in the *Index*. The *Manual* uses functional descriptions to group together things that "work" similarly or that achieve similar results (proximate function). By extension then, ask if a pencil could be defined as a "coating implement which applies a solid material through rubbing?" While this is a wordy way to describe a pencil, it is an accurate description. At the same time, this "title" also describes other objects, such as dispensers for solid anti-perspirant, which work on the same principle and would also be in this classification.

When using the *Manual* read the class title, the capitalized subclass titles, and any relevant indents as a sentence. After reading this sentence, ask if it describes your invention. Reading the *Manual* in this way will help ensure that you have not misapplied an indent to the wrong subclass. This type of misapplication is fairly easy, as figure 8.1 shows.

Fig. 8.1. Class 2 Apparel from the *Manual of Classification* (December 1986).

CLASS 2 APPAREL

DECEMBER 1986

BODY GARMENTS

.Shirts

115 ..Men's outer garments

116 ...With attached collars

117 ...With garment supporters

118 ...Bosom structure

119 ....Detachable

120 ....Supporters and protectors

121 ...Overlaps

122 ...Yokes, patches and facings

123 ...Cuffs

124 ....Detachable

125 ...Sleeves

126 ....Detachable

127 ...Neckbands

128 ...Closures

129 .Collars

130 ..Combined with neckties

131 ..Soft fold type

132 ...Supporters

133 ..Women's wear

134 ...Supporters

135 ..Waterproof

136 ..Paper

137 ..Necktie-engaging devices

138 ...Easy-slide structure

139 ..Attaching to shirts

Note that subclass 138, "Easy-slide structure," is under 137 "Necktie-engaging devices," which is then under 129, "Collars." Subclass 138 does not relate to subclass 136, "Paper," even though both are related to 129, "Collars." As a result, you would not expect to find patents for a sliding structure used to attach paper collars to shirts in subclass 138. This is because 138 should be read as: "Apparel, collars, necktie-engaging devices"; whereas 136 should read as: "Apparel, collars, paper."

While 98 percent of the *Manual* consists of class schedules, other useful information can also be found and used as additional access points. These other *Manual* sections include:

1. A complete tabulation of all classification numbers, both those currently being used and those no longer in use. Each entry gives the date of first usage and the date of the latest revision. This list can be scanned to determine if a class schedule is missing, or if it no longer exists. Generally, users get a "missing" class number from either an old patent, or, more likely, from the *Index* since the *Index* and the *Manual* can easily get out of sync because of their different publication cycles.

2. A list of classifications divided into the following "Main Groups":
   I. Chemical and Related Arts
   II. Communications, Designs, Radiant Energy, Weapons, Electrical and Related Arts
   III. Body Treatment and Care, Heating and Cooling, Material Handling and Treatment, Mechanical Manufacturing, Mechanical Power, Static and Related Arts

   These lists function as an organizational chart for the examiners and classifiers and include the following information for each group: name of director, group classifier, and heads of each art unit. Telephone numbers and room numbers are given for each. Although many users are reluctant to contact these people, a particularly good reason to contact them occurs when you cannot tell the difference between specific subclasses that have very different definitions, but that seem to house very similar patents.

3. A list of classes arranged by art unit. This is yet another way to find related classes and subclasses. An example of an art unit follows.

Art Unit 263,

		Subclasses From	To
Class 370	Multiplex Communications	All	
Class 375	Pulse of Digital Communications	All	
Class 455	Telecommunications	1	25
		31	620

Note that it is possible for subclasses to be in different art units. In 1987 subclasses from class 60, "Power Plants," were in art units 341, 342, 343, and 346. This is another demonstration of the interrelatedness of many aspects of technology.

4. A list of classifications in numerical order by class number giving class titles, subclasses assigned, and art unit. This list can be used to find out which art unit a specific class and subclass is assigned. The purpose of these lists is to let users have access to members of the Examiners' Corps assigned to particular art units.

5. A list of classes in alphabetical order by class title. This alphabetical list can also be used to supplement the *Index*, especially if you are having trouble finding or thinking of related index terms.

USING THE *CLASSIFICATION DEFINITIONS*

Classification Definitions (see appendix 18) are critical for determining the scope of a class because they describe the subject matter included *and* excluded. Consequently, the *Definitions* are the final authority on what subject matter will be found in any particular class and subclass. The *Definitions* state as explicitly and precisely as possible the boundaries of each class and subclass. The *Definitions* also draw the lines of distinction between classes and between subclasses within a single class. The *Definitions* often contain a glossary of terms and always contain references to other classes and subclasses housing related subject matter. These references are found under the headings: "Search Class; Search This Class, Subclass."

Returning to the pencil example, first locate either the microfiche card or paper copy of the definitions for class number 401. Read the class definition (known as the head note) for the entire class first. This gives an overview of the entire class and cross-references for other related classes. The class definition starts: "... class for manually manipulated device for applying or spreading a coating on a work surface." Ask yourself if this applies to your subject matter. If it does, move on and read the subclass definitions. Be sure to read all of the references relating to your class and subclass. It is quite common for a subclass definition to refer to another earlier subclass definition. All of these must be read and evaluated. For example, the definition for subclass 49 starts: "Implement under the class definition comprising a support adapted to utilize a piece of non-fluent, cohesive, self-sustaining coating material." Again, does this definition apply to your subject matter? Does it define pencil? Yes, a pencil is a support device for lead or graphite. Either can be defined as non-fluent, cohesive, self-sustaining coating material. The usefulness of the *Definitions* is even more clearly demonstrated by the following example.

Suppose that you are looking for patents that relate to toasting bread by using an electric grill type device. You look in the *Index* and discover that patents stored in Class 99, "Foods and beverages apparatus," subclass 385, "Slice Toaster or Broiler," seem highly relevant. The class definition for 99/385 states that: "This class comprises apparatus for preparing, treating, and preserving substances intended to be eaten and drunk by human beings or animals." The subclass definition further states that: "Cooking apparatus adapted to support food material in the form of a slice while undergoing heat treatment.... When the material treated is bread, the operation is known as toasting, and when meat, as broiling." The reader now knows that broiler units and toasters are essentially the same and that you would expect to find both in subclass 385. This is not how most of us think of broiling and toasting. Most of us, if asked, would probably have expected a difference. It is also interesting to note that both human and animal food treating devices are found in the same class: ovens for baking cookies, dog biscuits, and animal kibble are all classed together.

It may be appropriate to remind readers at this point that there are no definitions for the design classes. The only classes with definitions are the utility patents and plant patents. See appendix 18 for examples of both utility and plant definitions.

USING THE RETRIEVAL TOOLS*

Once appropriate classes and subclasses have been selected, it is time to retrieve the patent stored in the selected class/subclass combinations. Several tools are available to convert classes/subclasses to patent numbers.

U.S. Patent Classification—Subclass Listing. This is a microfilm publication listing all patents classified in each class/subclass produced by the Patent and Trademark Office. Patents, within each subclass, are listed in numerical sequence along with the reference and cross-reference designations OR, XR, and UX. This list is updated quarterly.

Index of Patents Part II—Subjects of Inventions. The *Index* Part II consists of the subject of invention of each patent issued, identified by the class/subclass numbers given the patent at the time of issuance by the Patent and Trademark Office. Only the classes and subclasses that have patents are listed. This can be a very difficult tool to use, because classes and subclasses are continuously revised, while this tool lists a patent only once, in the class/subclass assigned at the time of issuance.

CASSIS. CASSIS is an online and CD-ROM database used by the Patent and Trademark Office to list all patent numbers that have been assigned to a particular class/subclass as either OR, XR, or UX cross-references. CASSIS has several modes or ways to retrieve patent numbers; that used to retrieve patents stored in class/subclass order is called CM, or classification mode searching. CASSIS, available at all patent depository libraries (see appendix 13), is updated quarterly. CASSIS is discussed in more detail in chapter 10.

Official Gazette. Generally referred to as the *OG*, the *Official Gazette of the United States Patent and Trademark Office* has been published once a week since January 1872. The *OG* replaces the *Patent Office Reports*. The *OG* is published every Tuesday simultaneously with the publication of complete patents. *OG* consists of two parts: "Patents" and "Trademarks." The patent section is divided into three sections: "General and Mechanical," "Chemical," and "Electrical." The bibliographic data, largest claim, and relevant drawing are in numerical order within each section. The largest claim is defined as the one that encompasses the most, not the longest. The bibliographic data consist of title, patent number, filing dates, inventor(s), assignee name(s) at issuance, and the number of claims and drawings in the granted patent.

---

*Consult chapter 14 for a complete list of patent search tools listed in alphabetical order.

The other major sections of the *OG* include a list in numerical order of design patents, Patent and Trademark Office notices, and indexes. The indexes cover names of patentees, names of reissue patentees, names of plant patentees, and names of design patentees. Included with every name entry are assignees, patent number, classification numbers, and title of the patent. Patent and Trademark Office notices include government-held patents available for licensing, a list of certificates of correction, re-examinations, reissue applications (which unlike patent applications are open for public examination), patents, and a list of patent depository libraries including addresses and telephone numbers.

The *OG* can be used in a number of ways. The arrangement into broad subjects with the entries listed in class and subclass order permits easy reference to patents in specific areas. One of the major uses of the *OG* is business intelligence (finding out what competitors are doing).

The *OG* also provides a quick method for validating a particular field of search. This is especially useful when searching in a library that does not store complete patents in class/subclass order. Currently only the public collection in Sunnyvale, California, and the Public Search Room in Washington, D.C. store patents in class/subclass order. The other collections file patents in numerical order. Therefore, for all other PDL collections, once patent numbers stored in chosen classes/subclasses have been retrieved, the searcher must either examine the complete patents or the *OG* entry. In most collections complete patents are on microfilm and the *OG* is in book format. As a result, it is quicker for searchers to scan the *OG* to determine if they are on the right track. Always keep in mind a clear description of your invention because you are looking for the closest match.

One limitation to using the *OG* as a filter is that the bibliographic records do not include cross-reference classifications or field of search notes. These are only found on the face of modern patents (see chapter 3 for a discussion of patent documents). However, using the *OG* as a filter is faster than putting reels of microfilm on a reader.

After looking at the *OG* and determining that you are on the right track, it is time to examine the full patent. The scope of patent collections varies widely from library to library. Some collections have a complete run of all patents issued since 1872, while others have only recent years. These files are generally on microfilm, but some of the collections have paper copies of older patents. Holdings for the patent depository libraries are available through the *Directory of Patent Depository Libraries*, published by the Patent and Trademark Office and held by all patent depository libraries.

Select from the *OG* those patent numbers that seem most appropriate and examine the complete patent in the following order:

1. Flip to the first page of each drawing and examine it for relevance. While you cannot rely exclusively on the drawings to determine whether the patent being examined is either your invention or gives the information being sought, it is a good place to start.

2. Read the abstract.

3. When necessary, read the broadest independent claim.

4. Note down cross-references and field of search, checking for any pattern. Remember that classifications may have changed.

5. Check CASSIS either online or CD-ROM, or other online services for current class/subclass designations.

6. Note prior art patent numbers cited in the background section of the patent. Check them for relevancy in either the *OG* or the complete patents. If relevant, check *Manual* and *Definitions* of these new patents for class/subclass.

7. Throughout this process, continuously check for patterns and for patents close or parallel to your invention.

8. If a pattern or particular patent appears to confirm that you are on the right track, check that patent for current original and cross-reference class/subclass in CASSIS or another online service.

OTHER TOOLS

Classification Change Orders. Generally these are referred to as change orders and consist of notices that changes to the classification system and the definitions have been made. These result from reclassification projects undertaken by the Patent and Trademark Office to reflect new technologies. Each order consists of changes that will be made to the *Manual of Classification*, a list of subclasses established as a result of the reclassification, disposition of abolished subclasses, changes to the U.S. I.P.C. Concordance, and changes to the *Definitions*. (See appendix 27 for an example.) Change orders are most useful when an online search by class number says "invalid classification" and a quick check of the *Manual* shows your selected class to be active. This means the selected class or subclass is in the act of being revised. The change orders will show the disposition of the selected class.

Manual of Patent Examining Procedure (MPEP). This publication consists of a looseleaf manual that serves as a detailed reference on patent examining procedure for the Patent and Trademark Office's Examiners' Corps. Chapters of interest include 100, "Secrecy and Access"; 600, "Parts, Forms and Content of Applications"; 2100, "Patentability"; and Appendix R "Patent Rules."

Public Search Facilities. These consist of the Public Search Room and the Scientific Library located in the U.S. Patent and Trademark Office (Crystal Plaza, 2021 Jefferson Davis Highway, Arlington, Virginia, 20231, (703) 557-3158). Both are open to the public. The Public Search Room includes patents arranged in class and subclass order and CASSIS. The physical files of patents are called "shoes." Each of these shoes is supposed to have the most recently added patents on top. Because these shoes are arranged in random order, it is important to consult the "Directory of Class/Subclass Locations" in order to find the physical location of each shoe. The other collection is the Scientific Library, which is just that, a library of scientific and technical publications for use primarily by the examiners.

Patent Depository Library System. The dissemination of the technical information contained in patents is one of the primary missions of the Patent and Trademark Office. An important aspect of this mission is the designation of libraries around the United States as patent depository libraries (PDLs). These designated libraries receive microfilmed copies of current U.S. patents and are required to maintain at a minimum a twenty-year collection of earlier patents. These collections are open to the public and, in addition to having copies of patents, they provide access to CASSIS, the *Manual of Classification, Index to the Manual*, the *Classification Definitions*, the *Official Gazette*, and other relevant tools.

CONCLUSION

Subject searching is the most complex. It requires a knowledge of what can be patented, how patents are organized, and the search tools. Generally, when people talk about doing a patent search, they mean a subject-of-invention search.

The next chapter uses information discussed in chapter 3 on the parts of a patent, an understanding of the classification system gained in chapter 4, the steps of a subject search, and how to use the tools discussed in this chapter to work step-by-step through two sample searches.

9 Sample Patentability or Subject Searches

Is there any thing whereof it may be said, See, this is new? it hath been already of old time, which was before us.

—Ecclesiastes 1:10

Other chapters of this book have discussed the theory of patentability or subject searching, the organization of patents into classes and subclasses, the use of the search tools, and the nature of patents. This chapter applies this information to performing one subject-of-invention and one patentability search. (Hereafter both will be referred to as subject searches.) These sample searches were chosen specifically to illustrate the search process and therefore should help the reader consolidate all aspects of patents and patent searching discussed in previous chapters. The following searches demonstrate both the intellectual and procedural steps used in subject-of-invention and patentability searching. In the first search, the searcher knows a particular technical problem (subject-of-invention) has been solved and wants information on this solution. In the second search the searcher wants to find out if his idea has ever been patented (a patentability search). Both are performed using the same steps and tools; the only difference is in the results desired.

SEARCH PROBLEM 1

A patron came to the library and wanted to know who manufactures machines for making latex surgical gloves. This is a good example of subject-of-invention searching, since in many cases the potential searcher was not thinking of patents as a solution to his or her problem. While this search was done in a patent depository library using CASSIS, it could have been done on any of the commercial databases covered in chapter 10.

Step 1: The searchers looked in *Thomas Register* or other manufacturers' catalogs, such as the *Vendor Catalogs* from Information Handling Services, Inc. (IHS) to find the name of a manufacturer.

No luck, neither source listed any manufacturers of glove-making machines.

Step 2: The next step was to contact a surgical supply house selling latex gloves and ask for names of their suppliers of gloves or companies manufacturing gloves. The idea was to contact these glove-manufacturing companies and ask where they bought their machines. The surgical supply gave us three companies, and these were contacted.

No luck, none of these manufacturers would give out the names for "competitive" reasons.

Step 3: Three uses of the phrase "competitive reasons" rang a bell. We decided that maybe the process had been patented and that from the patent owner we could find out who had licensed his or her invention.

Step 4: We listed relevant, important, or operative parts of invention being sought.

a. A machine or process that makes latex gloves by dipping a mold into latex solution.

b. It also consists of some means for keeping the fingers from sticking together.

Step 5: We went to the *Index to the Manual of Classification*.

a. We looked up gloves.

b. We looked up dipping and found references to:

apparatus for molding	425	269
molds or forms	425	275
processes for forming by	264	301 +

Step 6: We went to the *Manual*.

a. We checked class 425 first, since it seemed most promising.

b. We next checked class 264.

c. We scanned schedule 425 in the *Manual*, reading the all caps until reaching: 425 PLASTIC ARTICLE OR EARTHENWARE SHAPING OR TREATING APPARATUS

269 DIPPING TYPE SHAPING MEANS

275 .Dipping or per se

We repeated this process with class 264 until reaching: 264 PLASTIC AND NONMETALLIC ARTICLE SHAPING OR TREATING PROCESS

239 MECHANICS SHAPING OR MOLDING TO FORM OR REFORM SHAPED ARTICLE

299 .Shaping against forming surface, e.g., casting, die shaping, etc.

301 ..accretion from bulk

306 ...conditioning or treating material or form to effect deposition

Step 7: We went to the *Definitions* and read the class descriptions found at the beginning.

a. We read the class definitions for 425 and 264.

b. We read the subclass definitions for 425/269 and 264/306.

c. We determined if the definitions described or fit the invention.

d. If not, we returned to *Index* or *Manual* and repeated the process.

e. If so, we continued to Step 8.

Step 8: We went to CASSIS or a commercial database.

The CASSIS commands are:

a. set pt on
This command tells the computer to print any patent titles in the database.

b. input: tmf 425/269
This command retrieves the title and indent level. Marked any XR references that looked particularly relevant.

Step 9: We searched selected OR and XR patents in the *Official Gazette*.

An examination of the OR references for 425/269 showed that class 425/269 was not relevant. However, several XR references were examined and patent number 4,521,365 was selected as being very "close."

Step 10: We returned to CASSIS or commercial database and got the original classification for 4,521,365.

The CASSIS printout looked like this in Patent Mode:

Patent selected: 4,521,365 - GLOVE APPARATUS AND METHOD

Number of clsfs: 6

Original classification: 264/306

1.	2/168	XR	4.	264/306	OR
2.	264/301	XR	5.	425/269	XR
3.	264/305	XR	6.	425/273	XR

We returned to Step 8:

Input OR or original class subclass 264/306 in CASSIS for the patents housed in this combination. Then went on to Step 11.

Step 11: We checked OR patents on the printout for relevance by reading the full patent.

a. We took special note of the "prior art" description.

b. We examined the disclosure section of the patent for clues or references to other related patents.

c. We examined the claims. (It is what is "claimed" that gives a patent its classification.)

Step 12: We went back to CASSIS (online or CD-ROM) or the commercial system and found the class/subclass.

a. For any cited patents in the disclosure section of the patent.

b. For any patent cited on the first page in the prior art references.

Step 13: We read full patents of those selected.

Step 14: It was time to make a decision.

Had we found a patent that solved our problem? Yes, patent 4,390,492, "Glove Molding Methods and Apparatus," granted to Leonard D. Kurtz and assigned to BioResearch Inc., Farmingdale, New York. The next step was to contact BioResearch for a list of companies licensing Mr. Kurtz's invention.

The glove search involved finding information stored in patents. It demonstrates how to do this type of search when you expect the answer to be in a patent and also serves as a reminder of what might be found in patents. The next search demonstrates a patentability search. The searcher had an idea and wanted to discover if it had been patented (he hoped not).

SEARCH PROBLEM 2

A patron had invented a new fabric that can be used as a closure or fastener. He had never seen anything like it and wanted to do a search. He called the Patent and Trademark Office or a local public library and was given the name of the nearest patent depository library (PDL). As a result, the following search was done in a PDL using the database available to PDLs, CASSIS. One of the advantages of searching in a PDL is direct user access to CASSIS. (For a further discussion of online searching using CASSIS and the commercial patent databases, please refer to chapter 10.)

Please note that while this particular search was done in a patent depository library (PDL) on CASSIS, it could have been done on any commercial patent database.

Step 1: He thought about what had been invented and listed the relevant, important, functional, or operative parts. (This can and should be done before beginning to search.) The relevant aspects of this invention are that:

a. It consists of two pieces of material.

b. The fastener effect is achieved through the use of two dissimilar materials which engage each other.

c. Engagement is caused by force or pressure applied to one or both of the pieces.

d. Disengagement occurs when a force is applied in opposite directions on one or both pieces.

Step 2: For this step the searcher needed to visit a PDL, or a library with the *Index to the Manual of Classification*. In the *Index* he looked up "fasteners" and other suggested terms and found the terms shown in figure 9.1.

Fig. 9.1. Results of search for "fasteners" in the *Index to the Manual of Classification*.

	CLASS	SUB
Fastener (See Connection; Joint; Key; Lock; Securing Means; Spline; Tie)	292	
Aircraft skin	244	132
Apparel	24	
Button and eyelet	227	
Coat collar	2	100
Fastener attaching	2	265+
Shoe sole	36	23*
Beds	5	498
Bolts	411	378*
Book binding	281	25R
Buckles buttons clasps etc.	24	
Clasp	24	455+

...

He scanned down the page until he reached the following:

Screws	411	378+
Separable-fastener, two part	24	572+
With protruding filaments	24	442+
Sheet fabric or web	24	
Slide	24	381+

...

He noted down these relevant/possible combinations:

1.	Buckles buttons clasps etc.	24	
2.	Composition fabric	269	906*
3.	Separable-fastener, two part	24	572+
	With protruding filaments	24	442+

Step 3: He went to the *Manual of Classification*.

a. He checked Class 24 first since it seemed promising.

b. He scanned the manual, reading all caps until reaching:

442 INCLUDING READILY DISSOCIABLE FASTENER HAVING NUMEROUS, PROTRUDING, UNITARY FILAMENTS, RANDOMLY INTERLOCKING, WITH AND SIMULTANEOUSLY MOVING TOWARDS MATING STRUCTURE

450 .Having opposed structure formed from distinct filaments of diverse shape to those mating

572 SEPARABLE FASTENER OR REQUIRED COMPONENT THEREOF

Class 24 subclass 442, class 24 subclass 450, and class 24 subclass 572 looked promising. A common, less confusing way to write class/subclass is 24/442 or 24/450. Because all three seemed to describe the invention quite well, he moved on to Step 4.

Step 4: He went to the *Classification Definitions* and read the class description (*Definitions* [Dates vary]).

a. He read the class definition for 24 BUCKLES, BUTTONS, CLASPS and found:

> This class provides for buckles, buttons, clasps, cord and rope holders, pins, separable fasteners, etc., which have become so varied in use and so allied in structure as to belong to no specific art, but are novel only as to their structures. There are, however, several types of fastenings included where the devices are but slightly identified with the art and are closely analogous to the main titles above cited. Such patents are retained under more or less art titles. Devices which embrace fastenings as above, but also include elements which connect them with various specific arts, have been excluded as far as practicable.

Since this looked good, he read the subclass definitions for 442, 450, and 572. He scanned down to 442 and found:

442. INCLUDING READILY DISSOCIABLE FASTENER HAVING NUMEROUS, PROTRUDING, UNITARY FILAMENTS RANDOMLY INTERLOCKING WITH, AND SIMULTANEOUSLY MOVING TOWARDS, MATING STRUCTURE.
Subject matter under class definition including means (1) for securing a segment of the structure-to-be-secured to either supporting structure thereof or a distinct segment thereof to a manner allowing the securement to be quickly detached, and (2) having a multiplicity of individual threadlike members which (a) each have all of their components integral with or fixedly attached to one another (b) are mounted to a common mounting surface anchored to the structure-to-be-secured or a support structure therefor from which they extend upwardly, and (c) are intended to move both without preorientation and in unison towards engagement with separate, opposed structure attached to, formed from, or consisting of the distinct segment of the structure-to-be-secured or the support structure therefore with which each threadlike member individually and mechanically interlocks.

(1) Note. To be proper for this and the indented subclasses several of the threadlike members should, by disclosure, be mounted along each side of the common mounting surface to allow random alignment of the means with the opposed structure at any given orientation of the mounting surface.

(2) Note. Patents which claim only structure details of a single, interlocking, threadlike member which is solely disclosed as being utilized in a securing means proper for this and the indented subclasses have been placed in these subclasses on a disclosure basis.

b. After reading the definition, he determined if it described the invention.

  1. If not, he returned to *Index* and repeated the process.
  2. If yes, he continued on to Step 5.

Step 5: He retrieved the patent numbers housed or resident in the chosen class/subclass combinations. Since this search was conducted in a PDL, CASSIS was used to retrieve the patent numbers. In most PDLs CASSIS can be searched directly by the user.

a. To retrieve patent number stored in a selected classification the CASSIS search command input is CM (classification mode) followed by the selected class/subclass combinations.

b. Prefacing a CM search with the command TMF (full title mode) results in the titles of the classes and subclasses being printed. This serves as a reminder of what was being searched and puts in perspective what was actually done. This is especially helpful if the search is carried out over a number of days. But more important, a TMF search acts as a check that the searcher has read the indent levels in the *Manual* correctly. This was done on the following search.

c. The CASSIS printout looked like figure 9.2.

Step 6: The next step was to scan the CASSIS printout.

a. He marked all the OR (original reference) references for examination.

b. He marked any XR (cross-reference) references that looked "particularly" good, noting that in 24/450 patent 4,307,493 used the phrase "velvet type fastener." While this is not how he would have originally described his invention, this description rang a bell.

Fig. 9.2. Results of search for selected classification in CASSIS.

Menu Mode-set pt on

Menu Mode-tmf 24/450

1.24 BUCKLES, BUTTONS, CLASPS, ETC.

2.24/442 INCLUDING READILY DISSOCIABLE FASTENER HAVING NUMEROUS, PROTRUDING, UNITARY FILAMENTS RANDOMLY INTERLOCKING...

3.24/450 .Having opposed structure formed from distinct filaments of diverse shape to those mating therewith

Clsf selected: 24/450

Number of patents: 51

1. 4,707,893 XR - title not yet available
2. 4,672,722 XR - Single Tape Closure Construction
3. 4,645,466 XR - Surfboard User's Foot Piece
4. 4,624,116 XR - Loop Pile Warp Knit, Weft Inserted Fabric
5. 4,488,335 XR - Hot Melt Adhesive Attachment Pad
6. 4,386,724 XR - Camera Strap
7. 4,386,723 XR - Firearm Sling Attachment
8. 4,344,654 XR - Wheel Trim Retention
9. 4,307,493 XR - Velvet Type Fastener Tape

[The OR patents first start to appear in this search at number 17.]

17. 3,863,304 OR - Linear Fastening Element and Method Thereof
20. 3,808,648 OR - Separable Fastening Sheet
21. 3,785,013 OR - Fastener
23. 3,708,837 OR - Fabric Fastener
24. 3,708,833 OR - Separable Fastening Device

[And on for 51 patents.]

Menu Mode-tmf 24/572

1. 24 BUCKLES, BUTTONS, CLASPS, ETC.
2. 24/572 SEPARABLE-FASTENER OR REQUIRED COMPONENT THEREOF

Clsf selected: 24/572

Number of patents: 64

1. 4,515,647 XR - Method and Apparatus for Forming an Integral Closure
2. 4,484,847 XR - Cargo Control Track and Fitting
3. 4,441,236 XR - Safety Lock
4. 4,402,114 OR - Snap Fastener Component Strip

[And on for 64 patents.]

Step 7: He searched the marked patents in the *Official Gazette (OG)*.

a. He read all abstracts or claims found in the *OG* and then characterized what was relevant or irrelevant about the selected class/subclass.

b. If necessary he rechecked the *Definitions*. (This would be necessary if a examination of the *OG* showed that he was not on the right track.)

c. If he seemed to be on the right track he continued to Step 8.

Step 8: He checked patents whose *OG* citations were relevant by reading the full patent.

a. He read the patent abstract.

b. He took note of the "prior art" description on the first page of the patent.

c. He checked field of search for other interesting class/subclass combinations.

d. He looked carefully at the patent disclosure section for references to other related patents.

e. He read all of the claims.

After looking up all the OR and many of the XR references in the *OG*, the searcher chose to examine the complete patents for 3,708,837, "Separable Fastening Device" and 3,708,833, "Fabric Fastener." This examination included reading the abstract and looking at the drawings, and because patent 3,708,833 seemed particularly relevant, the next step was to read the prior art or history section of the patent found in column 1 on the second page of the patent. Here the searcher found a *very* interesting sentence: "Separable fasteners such as those described in U.S. Pat. Nos. 2,717,437 and 3,009,235 which are marketed under the registered trademark VELCRO brand hook and loop fasteners."

Both of these references had to be checked; therefore, he went back to the *OG* or the full patents and read 2,717,437 and 3,009,235 for relevancy. They turned out to be highly relevant and in the actual search the patron stopped here. However, if the searcher had wanted to know more he or she would have continued with Step 9.

Step 9: Go back to CASSIS or a commercial system to find the class/subclass for the cited patents 2,717,437 and 3,009,235. In CASSIS the command input is PM (patent mode). The PM command retrieves original (OR) and cross-references (XR) for the selected patent numbers. In this case, the CASSIS printout looked like figure 9.3.

Step 10: Check the OR number, 428/92, in CASSIS or in the *Manual* for the title of class:

428	STOCK MATERIAL OR MISCELLANEOUS ARTICLES
428/85	PILE OR NAP TYPE SURFACE OF COMPONENT
428/92	.Particular shape or structure of pile

Step 11: Check the *Definitions* to discover the difference between 24/442 and 428/92.

Step 12: Make a decision as to whether or not the idea being searched has already been patented.

Have we found a patent for our idea? Yes, it is very likely patent 2,717,437, "Velvet Type Fabric and Method of Producing Same," issued October 22, 1951, to George De Mestral, Vaud, Switzerland, and assigned to Velcro S.A., claims priority. Or in other words, the searcher's idea of a fabric fastener has already been patented. What did we discover from this search?

Fig. 9.3. Results of search for specific patent number on CASSIS.

Patent selected: 2,717,437

Number of clsfs: 23

Original classification: 428/92

1.	2/DIG 6 XR	13.	128/DIG 15 XR
2.	2/235 UX	14.	139/391 XR
3.	2/265 UX	15.	160/DIG 7 XR
4.	2/321 UX	16.	242/74 UX
5.	5/496 XR	17.	273/DIG 30 XR
6.	5/498 XR	18.	273/346 XR
7.	15/159 A UX	19.	350/117 XR
8.	15/209 R UX	20.	428/86 XR
9.	24/445 XR	21.	428/92 OR
10.	28/161 XR	22.	428/95 XR
11.	55/DIG 7 XR	23.	446/901 XR
12.	55/DIG 31 XR		

1. That a product—Velcro™ brand fastener—already exists.
2. That improvements on this product also exist, particularly in regard to the "hooking" element.
3. That Velcro, Inc. has several related patents.
4. That many applications or uses of the original patents can be found, for example "velcro"™ used on diapers or in surgery.
5. That applications of this invention are generally found in the classification concerned with fasteners (24), while the first patent for material fasteners of the velvet type is in STOCK MATERIAL, class 428.
6. Examination of the claims shows why patent 2,717,437 was given an OR original classification 24/428. The claims for this patent discuss only the nature of the material; the possible application of this material as a fastener is listed in the disclosure. Remember, patents receive their original classification by their claims, not what is covered in their disclosures.

CONCLUSION

Obviously, the searches demonstrated here were edited, since in the real world of patent searching, each step could be repeated over and over and take hours, days, or even weeks. Accuracy is vital in each step and each iteration of the step.

While accurate retrieval is always important, in patentability searching accuracy is everything. Time and money are wasted if a search is poorly done and does not retrieve the desired information. Retrieval can be increased by taking time to analyze the problem before beginning to search, by using all of the tools carefully and intelligently, and by examining all relevant or "close" patents, no matter how old. Intelligent use of the tools will facilitate searching and is characterized by diligence: carefully reading the schedules, the definitions, and the patents. A mistake in reading the schedule titles and indents can lead to misleading results. It is also extremely important that all references in the *Definitions* to other classes and subclasses be followed back to the end. Attention to detail at every step improves retrieval, increases accuracy, and decreases the amount of time wasted searching in the wrong place. This same type of attention to detail must be used when examining full patents because of the somewhat arcane language. Therefore, it is wise to search in manageable amounts of time and to keep clear records of what has and has not been searched.

The best way to become familiar with difficulties that can be encountered using the search tools, in the search process, and in the language of patents is to do a number of searches yourself. The following problems are examples of typical subject searches; the answers can be found in chapter 13.

PROBLEMS

1. I have noticed that an increasing number of videotapes are copy protected in such a way that, while a copy can be made, it is unwatchable. Is this process patented?

2. Since 1983-1984, it has become very common in Texas, Florida, and other hot areas for cars in parking lots to be sporting sunscreens. These screens keep the steering wheel from becoming too hot to touch. They are used inside the car, fold up when not in use, and fit any car. The problem is to find the patent for this invention along with its class and subclass.

3. A number of companies are competing in the area of cardiac pacemakers. The problem is to discover what pacemaker advances Intermedics, Inc. has patented since 1983.

4. When the Soviet Union launched Sputnik it was a shock to the United States. Were there any satellite patents granted prior to Sputnik?

5. Find out who invented earmuffs.

6. When were combination smoke and heat detectors invented?

7. Computers have become commonplace in offices and homes. Earlier computers used a punched card system; was this system ever patented?

8. Has a machine been invented which can accurately focus a beam of light (weld) on the wires of computer chips?

Please select several of these problems and give them a try. If you have any difficulty go back to the section in the book covering that area and re-read it. Generally difficulty will center around misreading the *Manual of Classification* or not fully reading the *Classification Definitions*.

The next chapter covers specific aspects of online patent searching, including its strengths and weaknesses.

REFERENCES

Definitions to the Manual of Classification. [Dates vary] Washington, D.C.: U.S. Patent and Trademark Office.

Index to the U.S. Patent Classification. 1989. Washington, D.C.: U.S. Patent and Trademark Office.

Manual of Classification. [Dates vary] Washington, D.C.: U.S. Patent and Trademark Office.

10 Online Patent Information Systems

Now, what I want is, Facts ... Facts alone are wanted in life.

—Charles Dickens

Previous chapters have discussed in detail the major manual techniques and procedures used to retrieve information in patents or to discover if an idea has been patented. This chapter discusses using the computer to do speed searching and to do searches that were once nearly impossible. Before discussing the benefits of computer searching, it is important to cover what is meant by computer searching and define both the advantages and disadvantages of using computerized databases.

Over the last decade the computer has become an important time saver in retrieving pertinent references to conferences and journal articles. It is now very common for researchers to do a computer literature search prior to beginning a project, to update older information, or for a myriad of reasons. These same reasons, and of course validating an idea, have caused great interest among patent searchers in using the computer to remove, or at least overcome, some of the difficulties encountered in manual searching. It is clear that in the not-too-distant-future the computer will have a strong impact on patent searching. However, at this time searching for patent information online differs in a number of ways from other types of online searching. Generally, in the more technical fields the searchable record in the database consists of abstracts and titles of papers that are quite descriptive; this makes it possible to retrieve relevant citations through the use of keyword intersections along with assigned subject descriptors. The intent of online computer literature searching is to retrieve highly specific and tailored results with as few marginal, irrelevant citations, or "false drops," as possible. This is often referred to as high accuracy searching since the higher the accuracy the fewer numbers of "false drops." Rarely do searchers want or need to see every article no matter how remotely related. Even with these highly specific results, researchers often face a conundrum: they hope to find nothing, but by finding nothing they cannot be sure whether or not the job is really finished.

PROBLEMS WITH ONLINE PATENT SEARCHING

This problem is even greater when using the online patent files. In patent searching high recall is more important than highly tailored results. In fact, recall should approach 100 percent. In order to approach 100 percent recall, precision or accuracy is not a primary consideration. The searcher must accept that a certain amount of irrelevant matter (false drops) will be included because he or she is trying to ascertain whether a particular solution occurs in any of the nearly five million patents.

Other difficulties with online patent searching include short, often meaningless titles and abstracts. This is generally not a problem when searching technical databases, because it is rare for research articles to have vague or meaningless titles and abstracts. However, patent titles are notorious for their vagueness and lack of detail. Some common examples include "Electronic Device," "Motor-Pump Unit," or "Traction Vehicle." An even bigger problem is that the abstracts often leave a great deal to be desired and basically serve only to give a general overview of the patented item. As if vague titles and abstracts were not enough, the language used is often very abstract; for example, a toy flying saucer (such as a Frisbee™) is called "a levitating disk" and the terms "flying" or "saucer" are not used in either the abstract or title. This obviously causes problems, since keywords are only efficient and effective if they are present. Consequently, the use of formal descriptive language in patents means that the presence of what the searcher thinks of as meaningful keywords cannot be guaranteed.

A second problem area is that the patented invention may be in a different technology from that in which it is eventually applied. Take the example of a spool. It may have been invented to wind up rope and have been placed in a generic class. However, later inventions may use the spool to wind thread, cloth, or film. Another more modern example occurred in our sample patentability search. Here we discovered that the cloth fastener now known as Velcro™ was originally classed in a stock material classification, while later applications can be found in the classes covering medical and amusement devices. A keyword search of the commercial patent databases WPI, CLAIMS, or INPADOC retrieved the use of this material in applications involving surgery and diapers rather than the patent for the material itself. The original patent for Velcro can be found in appendix 28. Note that the abstract is a good example of formal descriptive language. It describes the invention but does not use just the keywords searchers might now associate with this invention. This patent is also an example of the previous style: it does not have an abstract, cited references are at the end rather than on the first pages, and there is no field of search information.

Two other areas of difficulty include extremely "new inventions" and state-of-the-art searches. Relevant references to very recent advances can be difficult to locate since the vocabulary may not yet be standardized or even exist. Searchers would need to think of related vocabulary and evaluate whether what they are looking for is the application of an earlier technology. For example, Kevlar™, an aramid fiber, was originally classed in 528, "Synthetic Resins," but it was patent 3,671,542, granted to Stephanie Kwolek for "Optically Anisotropic Aromatic Polyamide Dopes,"* that made commercial production possible of these fibers which would later be used to produce, among other things, tire cords and bullet-proof fabric. This means that searchers would not find patents for the fibers or the means to produce these fibers in classes dealing with tires or bullet proofing.

State-of-the-art searches can pose another type of difficulty. As in other searches, searchers must be careful to find the correct areas for examination. But the most common difficulty is that searches in active areas can turn up hundreds, even thousands, of references and each of these must be at least scanned. As a result, state-of-the-art searching can be extremely time consuming. For example, a search of class 364, "Electrical Computers and Data Processing Systems," subclass 900, "Miscellaneous Digital Data Processing," retrieved 5,033 patents in the database CLAIMS.

Even with these limitations, patent information is increasingly being sought through the use of computer databases.

As discussed, online patent searching has a number of peculiarities and inherent complexities different from those of other types of online information searching. One of the major difficulties is in choosing the correct database. For example, if searching for recent articles on laser bonding the searcher would generally have only one or two databases (COMPENDEX and INSPEC) to choose from. The task is more complex in patent searching because there are more databases to choose from and each differs somewhat in search capabilities. However, most are capable of searching and retrieving bibliographic data, classification information, and some type of surrogate such as an abstract or largest claim; they just do it in different

*U.S. patent 3,671,542 was such an important patent that Kwolek was granted the American Chemical Society Award for Creative Invention.

ways. Before discussing the characteristics and coverage of the patent databases, the next section describes basic types of online patent searches and, when relevant, the differences between the commercial databases and the Patent and Trademark Office database CASSIS are given.

BASIC SEARCHES AVAILABLE

Patent or Application Number. These are the simplest searches. All of the databases have a command which quickly brings up the record by country code and patent number. Patent number searches can also be used to determine the current class and subclass of any issued patent. This is particularly useful when, during the course of a search, you find a reference to an older patent number. A search by patent number on CASSIS online or CD-ROM or any of the commercial databases will retrieve the current class and subclass of any selected U.S. patent number.

Searches by application number are a bit more difficult. United States application numbers are secret until a patent is issued under that application number. However, several agreements have been made between patenting countries concerning application numbers under the PCT and WIPO treaties. Basically these agreements state that inventors must list all relevant application numbers received as a result of applying for patent protection in other countries. These U.S. application numbers will then be listed in any granted patent from participating countries such as United Kingdom, Japan, Germany, etc. An example and further discussion is found on page 74, "Ready mixed dry cement ..." U.S. application 447,937. This type of search is possible only on the commercial databases such as INPADOC, CLAIMS, and WPI.

Inventor/Patentee. These searches can be accomplished easily and quickly using any of the commercial online databases. They are nothing more than a basic author search and should be searched as such using the specific commands of the selected database. Generally the format of the inventor's name is highly controlled both by the issuing country and by the database. In the United States this control centers around including a middle initial; however, the inventor may choose to include his or her full middle name. This means that searchers should be prepared for name discrepancies. A search on CLAIMS (Dialog File 25) demonstrates some of these discrepancies. The search in figure 10.1 is for patents issued to the inventor of Kevlar™

Fig. 10.1. Inventor name search on CLAIMS on DIALOG File 25. Reprinted with permission of Dialog Information Services, Inc.

E AU=Kwolek S

E1	3	AU=Kwoleck J
E2	1	AU=Kwoleck John P
E3	1	AU=Kwolek S
E4	5	AU=Kwolek Stanley J
E5	2	AU=Kwolek Stephanie L
E6	9	AU=Kwolek Stephanie Louise
E7	1	AU=Kwolek Stephan S
E8	4	AU=Kwolek Stephenie Louise

Note E8 includes a misspelling of "Stephanie." A complete search would include looking at E3, E5, E6, and E8.

Assignee/Company. Coverage by company assignee varies considerably from database to database. Derwent's WPI and CASSIS CD-ROM assign a uniform company code to assignee firms which allows precise and easy searching. This is in contrast to the other databases, where the searcher must input possible variant names (e.g., Du Pont, Du Pont de Nemours). Retrieval is very easy if you have the code.

Unfortunately only the assignee designated at the time of issuance is listed. Searching by company name has all of the difficulties of variant names and subsidiaries which are found in online searching by sponsor, corporate source, or company name in more traditional bibliographic databases. It is not possible to search assignee or company name on CASSIS online and this capacity is limited on CASSIS CD-ROM to utility patents issued since 1969 and all others since 1977.

At this time, it is not possible to search online for the current assignee on either the commercial or the USPTO system. Current assignee must be traced manually through the inventor's name. In other words, the inventor must be located and asked if his or her patent is currently assigned (owned) by either a company or another individual.

Subject/Keyword. The two main access points for subject searching are patent classification and keywords. The effectiveness of keyword searching is directly proportional to the size of the record. A few of the databases, such as WPI, have added extended or rewritten abstracts in order to facilitate searching. Other databases make it possible to search both the claim and the abstract, and one includes the full text of the patent (see LEXPAT entry and discussion on page 79). However, all of the caveats mentioned previously should be considered before starting a keyword search. Keywords can sometimes be used effectively as an access to the classification if no relevant words can be found in the *Manual* or *Index*. This is especially effective with new technologies. Keyword searching is possible on CASSIS online and CD-ROM for only a limited number of years as storage space permits.

Classification Searching. Classification searching is the point of departure for all manual subject searches and can be very labor intensive and tedious. This search is used to determine which specific patent numbers are stored (listed) in a particular class/subclass combination. Any of the commercial databases and either CASSIS online or CD-ROM will speed up retrieval of patent numbers by class and subclass by printing a list all of patents resident, classified, or cross-referenced to a specific class and subclass since the first U.S. patent was issued.

Patent Families (Equivalents). This is one of the advantages of online searching because patent family searching is almost impossible to do manually. A patent family consists of all patent documents issued from patenting authorities on the same invention. These documents can include unexamined applications, examined applications, and granted patents from various countries. A patent family is created when an inventor obtains a patent protection for his invention in more than one country. These related patents are grouped into "families." The first patent entered in a database is designated as the "basic" patent. Any entered subsequently are known as "equivalent" patents and are generally not documented to the same degree as the first or basic patent. It is important to remember that the "basic" patent is only the first received by a database, not necessarily the first patent published or even the first application filed. It is only the first seen, and as a result the patent designated as the "basic" can differ from database to database.

Searchers should exercise caution, since there are a few patent families in which members do not protect the same aspects of an invention. This can be the result of different patent granting rules used by the issuing country, and generally arises when an application has multiple priority filings.

One of the uses of patent equivalents (family searching) is to find an English-language patent for a previously cited non-English-language patent. In the following example U.S. 4,464,201 is an English-language "equivalent" of Japanese patent JP 85-125646. Another important use involves U.S. application numbers. Since U.S. applications are secret prior to issuance of the patent, it may be desirable to find out if the application number was included in a priority filing for a foreign patent. The following example of an online family search demonstrates this relationship. United States application 447,937 was a "priority application number" in the Belgian patent BE 891434. The search in figure 10.2 was done on WPI.

Fig. 10.2. Patent Equivalency Search on WPI, DIALOG File 350. Reprinted with permission of Dialog Information Services, Inc. and Derwent Publications Ltd.

{Title} Ready mixed dry cement compsn. for use in reinforced concrete shell constructions.

Patent Assignee: (Cime-) Cimenteries CBR SA

Author (inventor): Pairon G S

Number of Patents: 008

Patent Family:

CC Number	Kind	Date	Week
BE 891434	A	820331	8216
EP 81861	A	820622	8226
JP 85125646	A	830726	8335
DK 82054433	A	830815	8339
US 4464201	A	840807	8434
ES 8402245	A	840416	8423
EP 81861	B	840919	8438
DE 3260757	A	841025	8444

Priority Data (CC,No, Date): BE 891434 (911210); BE 206801 (811210);

Applications (CC,No, Date): US 447937 (821208);

EP and/or WO Languages: French

This example shows that eight patents are related to each other and are said to constitute a patent family and are therefore "equivalent." Also given is the date and week each patent number was issued. Note that the priority or first patent issued in this group was a Belgian (BE) patent.

Another use of equivalency or family searching is to find the U.S. patent number for a cited U.S. application number. Our example shows that U.S. application 447,937 was later issued as U.S. patent 4,464,201. Application numbers can be cited in a number of sources, including *Chemical Abstracts*. This occurs because the inventor must state, at the time of application, any other applications which have been filed in other countries. It is very possible for an issued foreign patent to carry a U.S. application number since many countries issue patents more quickly than the United States. It is also possible to use online searching to determine if a cited application number has ever been issued as a U.S. patent since not all applications result in patents being issued—some applications are rejected for statutory reasons, being obvious, and of course, because the device, composition of matter, or process has already been patented.

Citation Searching. Patent citation searching is in many ways like searching SCISEARCH online or *Science Citation Index* manually. In either case, the idea is to search by known citation (patent number) to see if it has been referenced in a later patent. All modern patents carry references to earlier cited patents on the first page. These citations may be listed by the inventor as part of the application or by the examiner. This is often an effective way of locating information that has not responded to other approaches. Like SCISEARCH, the citations may turn out not to have as close a relationship as desired. Citation searching is impossible manually. One can search those cited by the patent in hand, but cannot search for those patents *citing* the patent in hand.

Citation searching can be very powerful; however, it is also very expensive. For example, each reference on CLAIMS costs $50.00. The following is an example. A patron heard on the radio that Arthur Fry invented the 3M product now known as "Post-it™ Brand Notes" about twenty years ago. The patron wants to know if this is true and if his patent was ever cited in a later patent. The following search was run as an inventor search in DIALOG File 23 (1950-1970) and 24 (1971-1981) CLAIMS, with the results shown in figure 10.3. (No results were found in File 24.)

Fig. 10.3. Patent Citation searching on DIALOG. Reprinted with permission of Dialog Information Services, Inc.

e au = FRY

Ref	Items	Index-term
E1	1	AU = FRUZZETTI PAUL R
E2	1	AU = FRY ARTHUR L
E3	5	AU = FRY ASHFORD B
E4	1	AU = FRY BERNARD ALBERT GERALS

s e4

S1 1 AU = "FRY ARTHUR L"

type s1/3/1

1/3/1

0315341 6630454

C/ ART MEDIUM

Inventors: FRY ARTHUR L (N/A)

Assignee: MINNESOTA MINING & MANUFACTURING CO Assignee CODES: 55992

	Patent Number	Issue Date
Patent:	US 3263605	660802

(Cited in 001 later patents)

Searching US 3,263,605 on DIALOG File 221, CLAIMS/CITATION revealed the following information:

Patent NO: US 3262605 (CITED IN 001 LATER PATENTS)

Cited Patents: US-02100358; US-02124294; US-02684012; US-02864882; US-02997453; US-03180260

Citing Patents: US-04279200.

To get more information on patent 4,279,200 we could look it up on one of the commercial online services or in the *Official Gazette*, or we could examine the complete patents. The complete patent is a better choice since we could verify which patents were cited on the first page of 4,279,200. In any event, the bibliographic information for our patent is as follows:

Method for Producing Nature Prints

Rosemarie Newcomb, 2227 Seely Dr., Orlando Fla. 32808.

Filed Dec. 26, 1979. Ser. No. 106,758.

U.S. CL. 101-170 6 Claims

Status Searching. Status searching is possible on several databases, including CASSIS CD-ROM and INPADOC. When searching CASSIS, the searcher is looking to see if specific patents have expired before their time for failure to pay maintenance fees or if they have been withdrawn. LEXPAT or INPADOC can be searched to determine whether any legal actions pertaining to that patent number have been reported. This type of search can be very useful to searchers who are looking for a technical solution to a problem, because any patent that has been withdrawn or has expired is now in the public domain and as a result, its solution can be used by anyone without infringement.

Status searching is also possible manually through the *Patent Status File™ Annual Cumulative Index 1973-1988*, published by Research Publications. Patents which have expired because of failure to pay fees can also be searched through a list available in all PDLs called *U.S. Patents Which Have Expired for Failure to Pay Maintenance Fees*. The latest update is May 15, 1989.

COMMON PATENT DATABASES

Table 10.1 and the following descriptions include only those databases which exclusively cover patents. Excluded are databases such as NTIS, CHEMICAL ABSTRACTS, RAPRA, PIRA, TULSA, and others that include patents along with journals, symposiums, conferences, etc. These were excluded from this discussion because patents represent a small number of the items abstracted and because search capabilities for the esoteric aspects of patents are limited. As a result, the following databases access only patents, their access tools, or legal decisions. If the database is available from one of the large vendors such as DIALOG or BRS, that is listed under vendor. If, however, the database is available only from the producer, such as CASSIS, the producer is listed as the vendor. WPI is a very special case, since only part of this database is available from a commercial vendor. The rest, which includes access to the special Derwent classification, is only available from the producer, Derwent. Consequently both producer and vendor are listed in the vendor column.

Table 10.1.

Coverage of the Patent Databases

Database Name	**Dates**	**Items**	**Vendor**
APIPAT	1964-	175,000	Pergamon
CASSIS	1790-	5,000,000	USPTO
CLAIMS-CHEM	1950-	1,700,000	Dialog
			Pergamon
CLAIMS/CITATION	1947-	3,500,000	Dialog
CLAIMS/CLASS	current	111,000	Dialog
			Pergamon
CLAIMS/COMPOUND Registry	current	14,000	Dialog
CLAIMS/U.S. PATENT Abstracts	1982-	250,000	Dialog
			Pergamon
CLAIMS/U.S. PATENT Abstract-weekly	1982-	no. issued	Dialog
			Pergamon
CLAIMS/UNITERM	1950-	1,600,000	Dialog
			Pergamon
EPAT	1978-	280,000	EPO
FPAT	1969-	580,000	Questel
INPADOC	1969-	14,000,000	Pergamon
JAPIO	1976-	2,000,000	Pergamon
LEXPAT	1975-	800,000	Mead
PATCLASS	1986-	4,500,000	Pergamon
PATDATA	1975-	700,000	BRS
PATDPA	1968-	1,500,000	Deutsches Patentamt
PATSEARCH	1971-	1,000,000	Pergamon
			InfoLine
USCLASS	1790-	5,000,000	Pergamon
U.S. PATENTS	1970	1,000,000	Pergamon
Full Text			
WORLD PATENT INDEX	1963-	3,500,000	Derwent
			Dialog
			Pergamon

Several of the following databases, including PATDATA and PATSEARCH, are produced from U.S. Patent and Trademark Office information. APIPAT, produced by the American Petroleum Institute, contains bibliographic data on U.S. and foreign patents of interest in petroleum refining, petrochemical, and other energy industries. CASSIS, produced by the Patent and Trademark Office, is available only through patent depository libraries or the Public Search Room of the Patent and Trademark Office, and consists of all U.S. patents arranged by class and subclass. Titles for utility for patents issued since 1969 and nonutility since 1977 are printed but are not searchable. Patent titles for current patents are not available until the file is updated, which happens every six months. Names of patentees from 1975- are available on CASSIS. Searchable fields include words from the *Index, Manual, Definitions*, and abstracts entered in 1986, along with numerical searches. Numerical searches include retrieval of all class/subclass combinations of a specific patent number and a list of all patents resident in a particular class/subclass.

CLAIMS/CHEM. Produced by IFI/Plenum Data Co., this database provides access to standard bibliographic information for U.S. chemical patents from 1950 and for electrical and mechanical patents since 1965. The files consist of two major segments, the chronological segments and the Uniterm series, and several related files, CLAIMS/REFERENCE and CLAIMS/CITATION.

CLAIMS/CITATION. Produced by IFI/Plenum Data Co., this database provides access to patents which are later cited by another patent. It includes citations to both U.S. and foreign patents.

CLAIMS/CLASS. Produced by IFI/Plenum Data Co., CLAIMS/CLASS provides access to the classification codes and titles for all classes and subclasses used by the U.S. Patent Classification System.

CLAIMS/COMPOUND REGISTRY. Produced by IFI/Plenum Data Co., this database consists of a nonbibliographic file of chemical compounds which appear in five or more patents. The file contains the IFI compound term number, compound name, available synonyms, molecular formula, fragment codes, and fragment terms. Each record is complete and does not represent a particular patent. This file is designed to be used in conjunction with Uniterm for locating patents related to compounds of interest. Compound term and fragment term lists are available for sale from IFI/Plenum.

CLAIMS/UNITERM. Produced by IFI/Plenum Data Co., this database provides subject indexing for chemical patents using a controlled vocabulary developed by IFI.

CLAIMS/U.S. PATENT ABSTRACTS. Produced by IFI/Plenum Data Co., this database includes bibliographic data and the patent abstract from the first page of a U.S. patent. It covers chemical patents since 1950, others starting in 1975. Design patents are covered since 1980.

EPAT. Produced by the European Patent Office, EPAT contains bibliographic and legal status for published European patent applications.

FPAT. Produced by the Institut National de la Propriete Industrielle, FPAT provides bibliographic access to French patents.

INPADOC. Produced by the International Patent Documentation Center, INPADOC covers all types of patent documents issued in forty-five national and international patent offices. Services include Patent Gazette, Patent Family Service, and Patent Register or Legal Status Service. The Legal Status file tells whether a patent has been granted and if it is still valid. Patent protection may lapse if the owner has not paid the fees.

JAPIO. Produced by the Japan Patent Information Organization, this database provides access to Japanese patent applications which were not covered by WPI. It includes English-language abstracts for most unexamined Japanese patents open for public inspection. Excluded are applications filed by foreign applicants, private Japanese citizens, and on technology peculiar to Japan.

LEXPAT. Produced by Mead Data Central, LEXPAT is designed to search and retrieve full-text patents.

PATCLASS. Produced by Pergamon InfoLine Inc., PATCLASS provides access to patent classifications including classes, subclasses, and traditional bibliographic data.

PATDATA. Produced by BRS, PATDATA contains citations, with abstracts to patents and all reissue patents and defense publications issued by the Patent and Trademark Office.

PATDPA. Produced by the Patent-und Gebrauchsmusterrolle des Deutsches Patentamts, this database contains German patent applications and patent citations.

PATLAW. Produced by the Bureau of National Affairs, this database covers the intellectual property decisions of the U.S. Federal Circuit courts and the Supreme Court from 1967 to the present. It is available from Dialog.

PATSEARCH. Produced by Pergamon InfoLine Inc., PATSEARCH contains bibliographic data for U.S. patents.

USCLASS. Produced by Pergamon/Orbit, USCLASS contains all U.S. classifications and cross-references for all patents issued since 1790.

U.S. PATENT OFFICE FILES. Produced by Pergamon/Orbit, this is a bibliographic file of all U.S. patents, continuations, divisionals, and defensive publications issued since 1970.

WORLD PATENTS INDEX (WPI/WPIL). Produced by Derwent Publications Ltd., WPI covers over three million inventions which are represented by six million patent documents from thirty patent issuing agencies. It also provides access to equivalent patents grouped together by patent families.

The primary patent databases, used most frequently because they are the most flexible, are:

Derwent's WORLD PATENTS INDEX

IFI/Plenum's CLAIMS

JAPIO's PATENT ABSTRACTS of Japan

Derwent's U.S. PATENTS

Mead's LEXPAT

INPADOC

The next section divides these most frequently used commercial databases into types of searches.

SUMMARY OF DATABASES BY SEARCH TYPE

Subject Matter

WPI and WPIL (ORBIT); and 350 and 351 (DIALOG)

JAPIO (ORBIT)

CLAIMS (ORBIT); 340 and 125 (DIALOG)

USPM or USPA and USPB (ORBIT)

English Equivalent and Patent Families

WPI and WPIL (ORBIT); 350 and 351 (DIALOG)

INPADOC and INPANEW (ORBIT); 345 (DIALOG)

Citation Searching

USPM or USPA and USPB (ORBIT)

WPI and WPIL; 350 and 351 (citations from EP and WO only)

LEXPAT (MEAD)

Assignee/Inventor

CLAIMS (ORBIT); 340 and 125 (DIALOG)

INPADOC and INPANEW (ORBIT); 345 (DIALOG)

USPM or USPA and USPB (ORBIT)

WPI and WPIL (ORBIT); 350 and 351 (DIALOG) (inventor information not complete)

JAPIO (ORBIT)

LEXPAT (MEAD)

Class

CASSIS (USPTO)

CLAIMS (ORBIT); 340 and 125 (DIALOG)

USPM or USPA (ORBIT)

WPI and WPIL (ORBIT); 350 and 351 (DIALOG)

INPADOC (ORBIT); 345 (DIALOG)

JAPIO (ORBIT) [International Classification only]

LEXPAT (MEAD)

CONCLUSION

It should be remembered that some of these online databases have paper equivalents and that most have their beginnings in tools produced by the respective national patent offices. The role of the commercial databases has been to take these paper equivalents and the computerized files of the national offices and disseminate them to a wider audience, at the same time enhancing them through the addition of extended abstracts, citation searching, and chemical compound registers. For example, WPI has added extended titles and abstracts to the database, while CLAIMS allows keyword searching in both the abstract and claim.

It is well to remember that online patent searching is in its infancy, and in all likelihood many of the difficulties encountered with keywords or vague titles will be fixed in the future as major changes are made. Some of the expected changes will include advances in full-text searching and the ability to search the drawings directly. Even with the current limitations discussed in this chapter, there are a number of searches that are only possible online, such as patent family and patent citation searching, and some such as patentee and assignee are made faster and less tedious.

Last but not least, online searching is another access point to the all-important classification system. It can be used to find a starting place when the searcher is unable to find a relevant class/subclass in *Index to the U.S. Patent Classification* and the *Manual of Classification* and it can be used effectively to list the class/subclass of any known patent or to retrieve an accurate, up-to-date list of patents classed in a specific class/subclass.

The starting points for any online patent search are:

buzz words, also known as keywords

company names

inventor names

patent number

patent classification numbers

Up to this point only cursory mention has been made in this text of foreign patents; the next chapter provides more information on this topic.

PROBLEMS

1. A package of swim goggles lists two patent numbers, U.S. D 280,670 and U.K. 1015921. What does each protect? And does each protect the same thing?

ADDITIONAL READINGS

Bechtel, H., et al. 1985. "Online Patent Searching—Useful, but Still in Its Infancy." *World Patent Information* 7, 1/2: 68-82.

Eisengschitz, T. S., and J. A. Crane. 1986. "Patent Searching Using Classifications and Keywords." *World Patent Information* 8, 1: 38-40.

Kaback, S. M. 1987. "Crossfire Patent Searching: A Dream Coming True." *Database* (October): 17-30.

______. 1983. "Online Patent Searching: The Realities." *Online* O (July): 22-31.

11 Foreign Patents

In every matter that relates to invention, to use, or beauty, or form, we are borrowers.
—Wendell Phillips

From time to time searchers performing state-of-the-art, patentability, and subject-of-invention searches will need access to foreign patents. The searchers may, during the course of a search, find a reference to a foreign patent or a reference to a priority filing that includes a U.S. application number. These are the most common, but other reasons for needing access to foreign patents include the need for information on the activities of foreign competitors, the need to be sure that an idea has not been patented in another country, and the suspicion that a patent is being infringed upon.

This chapter introduces foreign patents rather than discussing foreign patent law or how to apply for a patent outside the United States. It is intended to serve as an overview, primarily so that searchers will not be uncomfortable should they run across a citation to a foreign patent.

INTERNATIONAL PROTECTION

Almost all countries provide some form of patent protection. However, a wide variety of patenting protection procedures and practices are found around the world in spite of a number of efforts, such as the Patent Cooperation Treaty (PCT) and the European Patent Convention (EPC), to standardize many aspects of patent documentation and application procedures. Some of these differences include what may be patented; the length of time for which protection is granted; whether or not the application is examined for patentability before being issued; the numbering system; and the procedures for establishing filing, publication, or priority dates. See appendix 29 for an overview from the Patent and Trademark Office of these major differences.

Currently over a hundred countries publish some type of patent document or specification. These documents consist of two basic types, unexamined and examined. Unexamined patents, often called applications, are issued without any attempt to search the "prior art." In these countries, the issuing office checks only for procedural defects, incomplete bibliographic information, and clearness of the drawing(s) before issuing the document. At present a majority of patenting countries, including Great Britain and Japan, publish unexamined applications and are often referred to as "quick patenting countries," because they can issue a document often twelve to fourteen months faster than countries that examine each application. These unexamined applications carry all patent rights until someone takes issue with them and asks for a prior art examination. As a result of this requested examination, the patent application can be validated and issued as an examined patent, it can be totally rejected with all patent rights abrogated, or it can be issued after changes have been made.

U.S. AND OTHER PATENT SYSTEMS COMPARED

In the United States a patent is issued or published only after it has been examined. The official procedures used in this process are discussed in chapter 12, "The Patent and Trademark Office."

Both examined and unexamined patent documents are country specific and provide protection only in the issuing country. The extent of this protection is determined solely by the national laws of each country. Therefore, U.S. patent protection is enforceable only in the United States, its territories, and its possessions and Japanese protection is good only in Japan, etc. However, whether issued as an unexamined application or as an examined patent, each type has the inventor's name, patent number, title, date, and some type of classification. At present over forty-seven countries classify published patent documents using the International Patent Classification System (IPCS). The IPCS differs substantially from the classification system used in the United States. However, since 1969 the United States has included the IPCS number as part of the bibliographic description. It is found in field [51] on the first page of the patent. The reverse is not true: the U.S. classification number is not printed on patents issued in other countries. A concordance to the two systems is published by the Patent and Trademark Office. *Concordance—United States Patent Classification to International Patent Classification* (available from the Patent and Trademark Office or in any PDL) relates U.S. classification numbers to general ranges of IPC numbers. A complete one-to-one relationship cannot be attained because of the differences in philosophies between the two systems, even where a single IPC notation is listed for a single U.S. subclass. As a result, an IPC listing can only be used as a guide to a field of adjacent groups. In all countries, printed specifications (patents), patent applications, and official patent journals and gazettes are the source material for determining whether a specific invention has been disclosed in some earlier foreign or domestic patent. Other than priority searching (determining if something has already been patented) there are other reasons for retrieving a patent issued in another country.

REASONS FOR SEARCHING FOR FOREIGN PATENTS

One of the more common occasions on which a searcher will want to retrieve a foreign patent is when he or she gets a citation to a foreign patent as the result of an online search for articles or conference papers. Here the searcher is generally not specifically looking for patents, but comes across them in the course of doing a search. This is most apt to happen when searching either *Chemical Abstracts* or *Petroleum Abstracts*. Both index U.S. and foreign patents.* It is important to remember that inventors with ideas that are economically valuable often file for patent protection in several countries simultaneously, and in fact, in order to be sure of receiving patent protection, it is important to file within the same twelve months in every country where protection may be desired.

Because of the long lag time between application and issuance in the United States it is possible that *Chemical Abstracts* will index a German or Japanese patent application for an invention that will later be patented in the United States. This happens because *Chemical Abstracts* always indexes the first patent seen, and it is often a foreign patent.** The process of converting a foreign patent number to a U.S. number is covered in chapter 10. One reason for this lag time is that application countries, because they do not examine each patent for novelty, obviousness, or newness, are able to issue faster. As a result, it is very common for a foreign patent application to be issued months prior to a patent being granted in the United States. As a result of patent treaties it is common for these foreign patent applications to include a U.S. application number. These treaties require inventors to list any outstanding applications on every other application.

---

*Both *Chemical Abstracts* and *Petroleum Abstracts* sell handbooks and/or workbooks which describe the mechanics of conducting a literature search, including searches which include patents, which specifically exclude patents, and searches of those patents indexed as part of their service.

**Examples: PCT WO 89 08,092, issued to Kevin Peter Wainwright for "Removal of organic compounds from fluids"; U.S.S.R. SU 1,502,592 issued to V. D. Koshevar et al. for "Protective acrylate coatings for anti-glare transparent screens."

Other reasons for needing foreign patents include finding a reference in a U.S. patent, keeping up with foreign competition, and watching for patents that might infringe on an invention.

Patents are, of course, published in the language of the issuing country unless they have also been filed under the EPC or PCT. Under the EPC or PCT, patents may be filed in the language of the country of any of the signers. Since Great Britain is a signer, patents filed under EPC or PCT may be prepared and prosecuted in English. Not only does this save the inventor money, but it also means that it is sometimes possible to convert a non-English-language patent citation to an equivalent English-language application or U.S. patent. For example, if the citation is to a German patent it may be possible to find an equivalent patent issued in Great Britain. The actual conversion process is discussed in chapter 10.

Copies of many foreign patents may be purchased from the Patent and Trademark Office or directly from the national patent office publishing the patent. Selected foreign patents are also available from a number of commercial document delivery companies. One of the better known is Rapid Patent in Washington, D.C. A number of patent depository libraries (PDLs) also receive selected foreign patents.

Readers may also want to know the basic steps inventors should use in determining whether to seek patent protection outside of the United States:

1. Decide where patent protection is actually needed:
   - —where the inventor is likely to make or sell the invention
   - —where he is likely to license his invention
   - —where his competitors are located.

2. Compare the disadvantages and advantages of filing:
   - —in each individual country
   - —under European Patent Convention (EPC)
   - —under Patent Cooperation Treaty (PCT).

3. Compare costs of each filing avenue (PCT, EPC, etc.).

The number of countries providing patent protection and the variety of protection available presents problems for both the inventor and the patent searcher. It is important that inventors and searchers take the time to plan and think so as to maximize search time and protection.

Chapter 12 provides an overview of what happens to a patent application as it goes through the Patent and Trademark Office.

ADDITIONAL READINGS

Grubb, P. W. 1982. *Patents for Chemists*. Oxford: Clarendon Press.

Patents throughout the World. 1982- . New York: Clark Boardman.
This loose-leaf service contains an overview of the protection and application requirements for most patent-issuing countries.

12 The Patent and Trademark Office

A country without a patent office and good patent laws was just a crab, and couldn't travel any way but sideways or backways.

—Mark Twain

Since this book has been about the nature of patents and how to conduct a patent search, especially a prior art search, it now seems appropriate to include a short explanation of what actually happens to a patent application once it reaches the Patent and Trademark Office. Following is an overview of the organization of the office.

BACKGROUND

Originally, during George Washington's administration, the Patent Office was one of several functions of the Department of State. The work was done by one clerk until the position of Commissioner was created by the Patent Act of 1836. Prior to 1836, the head of this office was the Superintendent. In 1925, by executive order, the Patent Office was transferred to the Department of Commerce, where it remains today. Now officially known as the United States Patent and Trademark Office or USPTO, it is headed by the Commissioner, who is appointed by the president and has responsibility for reviewing patent applications and determining whether or not to grant a patent. The office is not responsible for enforcing or adjudicating either patent rights or infringement; this is handled through the federal court system. Within the USPTO, the responsibility for determining patentability and conducting prior art searches rests with the patent examiners, who also check each application for ambiguity, defects in the drawings, and violation of the rules. These examiners are organized into art groups (subject areas), with each examiner responsible for "examining" or evaluating applications in his or her subject specialty or area of expertise. At the present, the Examiners' Corps consists of over 1,400 scientists and engineers, most with advanced degrees. Even with this extensive technical training, examiners go through additional training in patenting procedures, theory, classification, and prior art searching after joining the office. This training is as follows (*MPEP* 1989):

1. The Patent Academy consists of initial training, and refresher seminars.

2. Legal Training consists of in-house legal courses; legal lectures on topics pertinent to examination.

3. Law School Program consists of formal after hours law school courses.

4. Technical Training consists of in-house courses covering: computers, electronics, hydraulics, and biotechnology.

5. Technical Course Program consists of formal after hours university courses.

6. Other Training in areas such as: supervision, and attendance at technical meetings and seminars.

7. Examiner Education Program consists of visits to technical facilities to gain first hand experience with the art they examine (paid for by outside contributions).

After extensive academic technical training along with in-house training, the examiner is ready to analyze and evaluate patent applications in his or her area of expertise.

THE PATENT PROCESS

Following are the steps for handling an application after it reaches the office:

1. Receipt of the application and pre-examination processing (see figure 12.1)
   a. This consists of establishing a record, known as the file wrapper. The file wrapper will eventually hold all documents and correspondence.
   b. During this process each application is checked for formalities (e.g., inclusion of oath, fees, etc.).
   c. The application is assigned to an art group.
   d. The art group schedules or dockets the application for examination.

2. Examination procedures (see figure 12.2)
   a. File is assigned to an individual examiner.
   b. The examiner reviews the application.
   c. The examiner conducts a prior art search.
   d. Examiner evaluates application for patentability, including whether:
      1. The subject matter is patentable.
      2. It meets the utility test.
      3. The application gives complete disclosure of how the invention works.
      4. It meets the novelty test.
      5. The disclosed is unobvious "to one skilled in the art."
   e. Patent is granted (see no. 3) or rejected.
   f. If application is rejected, a response is sent to the applicant or his or her agent stating the reasons for rejection. Suggestions for changes which might make it possible to get a patent may also be sent.
   g. A response to the rejection is received from the applicant or his or her agent and the office issues a final rejection:
      1. At which time the applicant abandons the application.
      2. The applicant appeals his rejection.
      3. In which case the appeal may be rejected, sent for reexamination, or granted.

3. Post examination
   a. Quality review designed to insure high quality in granted patents.
   b. Determine issue fee.
   c. Authorized printing of the *OG* and full patent.
   d. Patent issues.

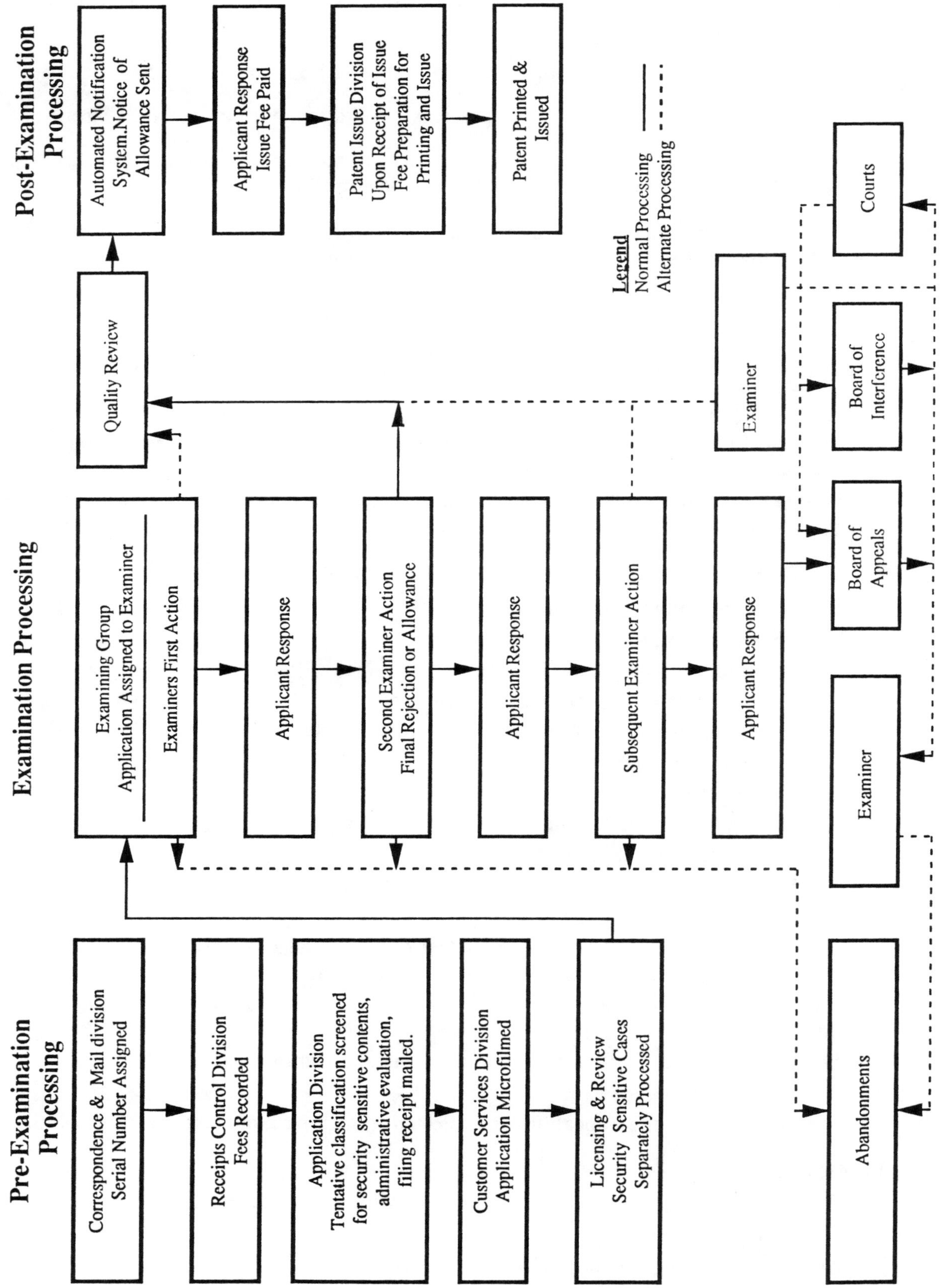

Fig. 12.1. The patent process. From U.S. Patent and Trademark Office.

Pre-Examination Processing

Correspondence & Mail division
Serial Number Assigned

Receipts Control Division
Fees Recorded

Application Division
Tentative classification screened for security sensitive contents, administrative evaluation, filing receipt mailed.

Customer Services Division
Application Microfilmed

Licensing & Review
Security Sensitive Cases
Separately Processed

Examination Processing

Examining Group
Application Assigned to Examiner

Examiners First Action

Applicant Response

Second Examiner Action
Final Rejection or Allowance

Applicant Response

Subsequent Examiner Action

Applicant Response

Quality Review

Examiner

Post-Examination Processing

Automated Notification System.Notice of Allowance Sent

Applicant Response
Issue Fee Paid

Patent Issue Division
Upon Receipt of Issue Fee Preparation for Printing and Issue

Patent Printed & Issued

Legend
Normal Processing
Alternate Processing

Abandonments

Examiner

Board of Appeals

Board of Interference

Courts

Fig. 12.2. Patent examining activities. From U.S. Patent and Trademark Office annual report (1989).

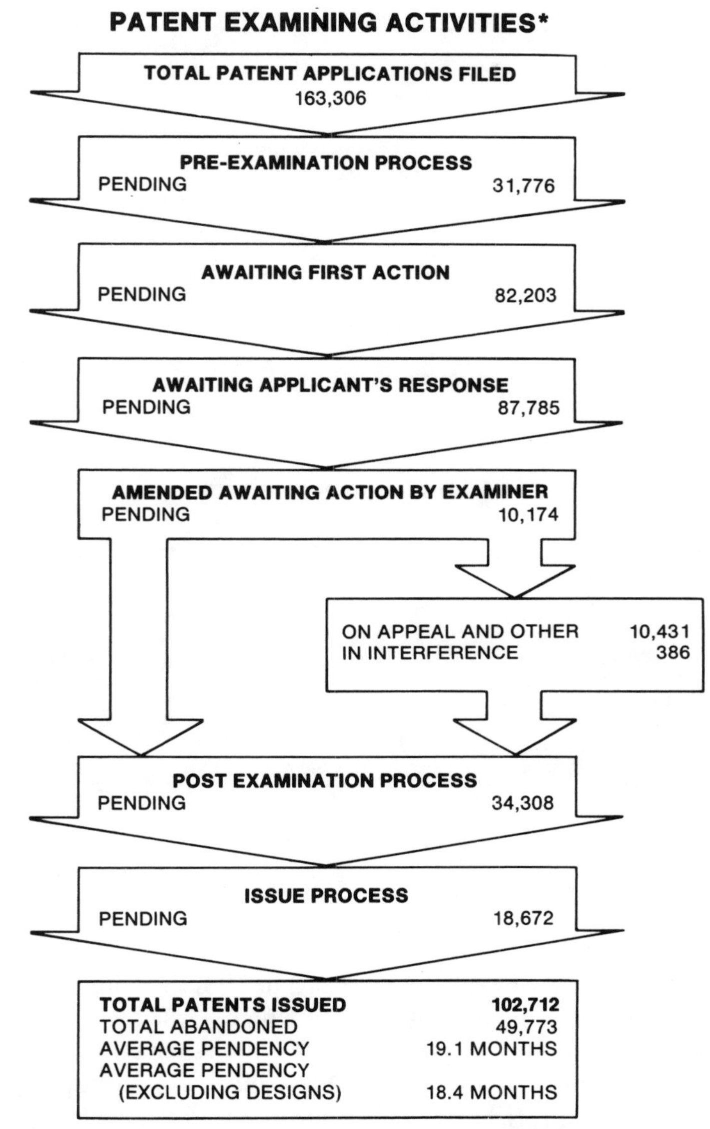

Over the years there has been a lot of public discussion about why it takes so long for a patent to be issued. There are three basic reasons: each year more and more applications are received (see figure 12.3), the increasing complexity of technology, and the sheer amount of prior art which must be examined. The following data demonstrate the size of the problem. The U.S. examiner's file now consists of 27,600,000 U.S. and foreign patents with a combined total of over 214,000,000 pages. A typical subclass has 240 documents of which 123 are U.S. patents, 106 are foreign patents, and 11 are miscellaneous (Quigg 1989). The average modern U.S. patent has 5.5 pages of text and 2.75 pages of drawings.

Fig. 12.3. Patent applications filed, pending, and issued (1969-1987). From United States Patent and Trademark Office annual report (1989).

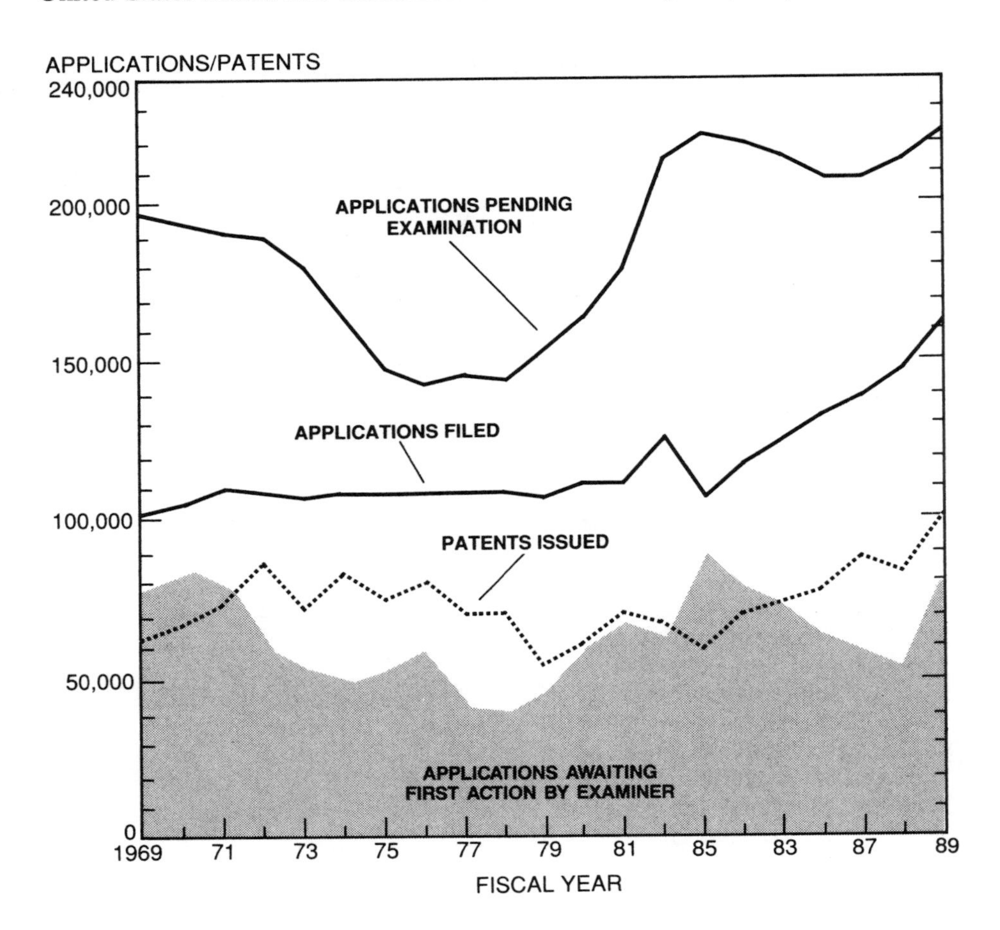

In 1988 a record 148,183 utility, plant, design, and reissue patent applications were filed. This represents an increase of 11,010 applications over those filed in 1987. During the same period, fiscal year 1988, 83,584 utility, plant, and design patents were issued, of which 77,884 were utility and 5,740 were design. On top of this, the office estimates that 1.8 percent of the patents in their files are reclassed annually (Department of Commerce 1989). It is no wonder that processing an application and granting a patent is such a lengthy process. These data reflect only the physical size of the file, they do not take into consideration the increasing complexity of technology or emerging technologies.

Increasing complexity and emerging technologies can best be demonstrated by the amount of reclassification done each year. Reclassifications are undertaken to increase the access to, and the reliability of, the patent search file. In 1988 the office established 2,950 new subclasses, which involved nearly 363,000 patents (Department of Commerce 1989).

To cope with this enormous increase, the Patent and Trademark Office is engaged in a massive automation project that is intended to help speed up the process. Still, issuing patents will continue to be a lengthy and time-consuming process, because even with automation all patent applications will still need to be read, studied, and analyzed by the examiners before the application can be compared and contrasted with relevant prior art. It is important to remember that when the patent system was first established the United States had less than four million people and its westernmost states were Pennsylvania and Virginia.

The intent of this brief outline is to give the reader an overview of the complex organization of the PTO. It is not intended to be exhaustive. (See also figure 12.4.) Along with the rest of this book, it is

Fig. 12.4. Organization of U.S. Patent and Trademark Office.

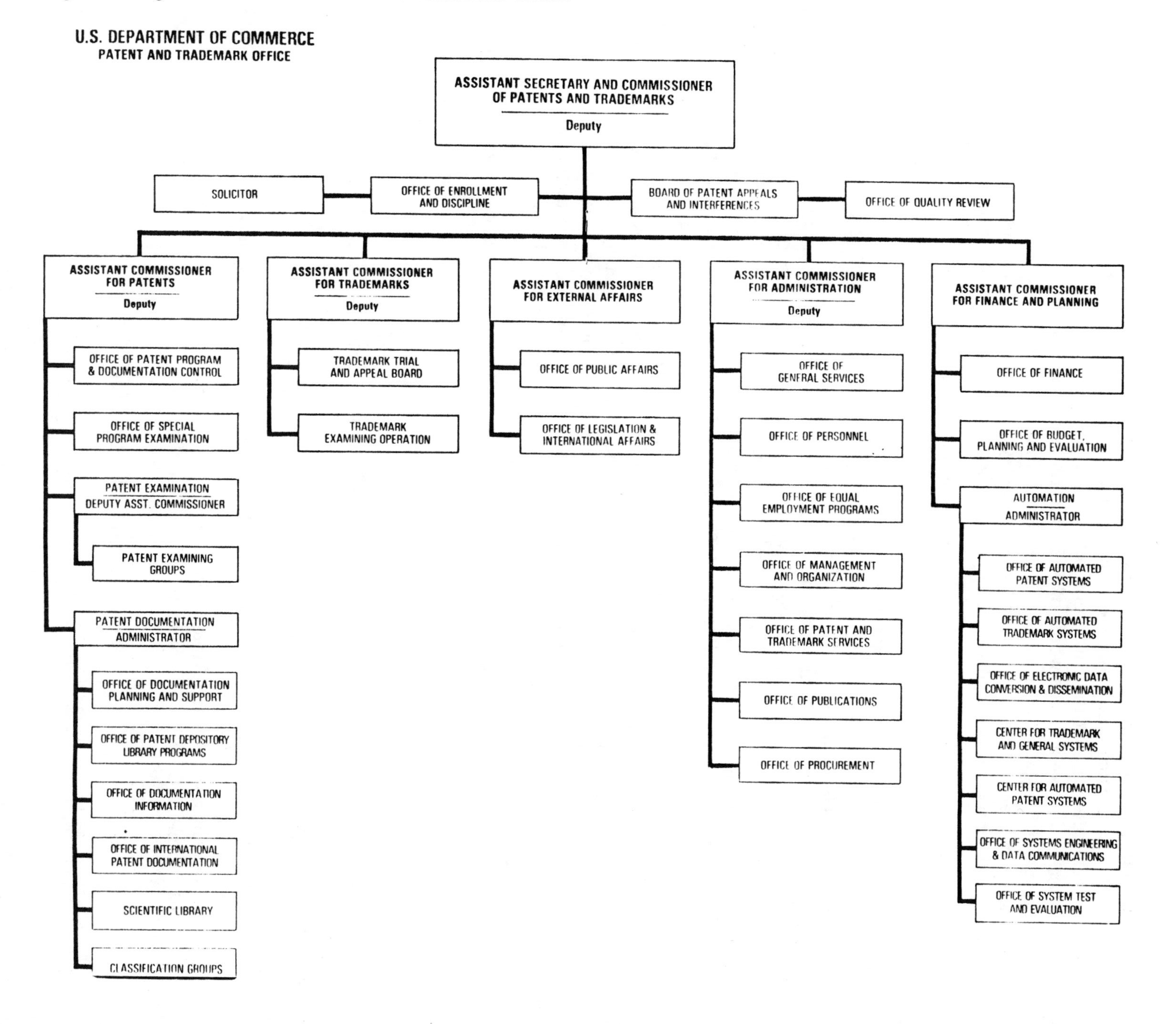

intended to increase the reader's understanding of the Patent Office's organization and add to an understanding of what is patentable, how patents are organized, and the techniques used in searching for patents. It should help ease some of the frustration often felt by searchers and inventors.

An inscription based on the words of Abraham Lincoln found above the northernmost portico of the 1936 Patent and Trademark Office building sums up the intent and goal of the U.S. Patent System and why the office encourages dissemination and use of this system by citizens whether they are corporate or individual inventors: "The patent system added the fuel of interest to the spark of genius."*

CONCLUSION

Genius and the will to invent are part of the human experience and invention is as old as the human race. Our forebears contributed many inventions, such as the irrigation ditches, writing, the violin, and the wheel. These inventions changed people's lives.

Quoting Donald Quigg, former commissioner, U.S. Patent and Trademark Office: "We do not often stop to think from where our blessings come. We accept them with little sense of gratitude. But in the light of our reflections, we can realize how vast is the obligation which we owe to the inventors of America. Not a meal we eat, not a paper we read or a television we watch, not a tool we use, not a journey we take but makes us a debtor to some inventor" (Quigg 1989). Today's inventors, like inventors in the past, must have certain characteristics, including imagination, enthusiasm, curiosity, perseverance, the urge to make a contribution, creativity, and the desire for financial rewards. However, unlike inventors in the past, today's inventors must also have patent-mindedness and education. Inventors must be aware of the importance of patents, the rules for obtaining a patent, and the techniques for using patents, and as inventions become more complex and complicated, formal education becomes more and more important.

The goal of this book is to encourage patent-mindedness by providing information on what can be patented; knowledge of and experience with patent organization; and practical tips for performing known inventor, state-of-the-art, and patentability searches. The more a searcher knows about the process, its idiosyncrasies, its strengths, and its weaknesses, the better he or she will be able to extract the information stored in patents.

With time, patience, and knowledge anyone can perform the basic searches described in this book and avoid the embarrassing situation of reinventing the wheel.

REFERENCES

Quigg, D. 1989. *Report to the American Bar Association. Section of Patent, Trademark and Copyright Law*. Washington, D.C.: U.S. Patent and Trademark Office.

Manual of Patent Examining Procedure. 1989. Washington, D.C.: U.S. Patent and Trademark Office.

U.S. Department of Commerce. 1989. *Commissioner of Patents and Trademarks Annual Report-1988*. Washington, D.C.: U.S. Patent and Trademark Office.

ADDITIONAL READINGS

Brewer, G. E. F. 1985. "The Seven Stages of an Invention." *Journal of Chemical Technology* 57, 723 (April): 55-57.

Iandiorio, J. S. 1988. "Patent Examiner." *IEEE Technology and Society Magazine* 7, 2 (September): 11-12.

"Is the Day of the Patent Over?" 1981. *New Scientist* 110 (September): 653-55.

---

*Lecture delivered to Phi Alpha Society of Illinois College, Jacksonville, Illinois on February 11, 1859.

13 Answers to Problems

Creativeness often consists of merely turning up what is already there. Did you know that right and left shoes were thought up only a little more than a century ago?

—Bernice Fitz-Gibbon

This chapter consists of answers to the problems found at the end of Chapters 6, 7, 9, and 10.

CHAPTER 6

1. The number D 282,087 was found on the underside of a small car. Is this a patent number, from what country, and what does it protect?

D 282,087 is a U.S. design patent. Design patents can be found at the end of each week's *Official Gazette* in numerical order. They can also be found, in many PDLs, by looking the number up on the Design Patents microfilm set. This number protects "Reconfigurable Toy Vehicle" issued to Ohno Kouzin on October 22, 1985 and covers a design for a car that transforms into a robot-like figure. Design patents protect the ornamental aspects of this invention, or how the car and the robot look.

2. Is 1,090,018, a number found on box at an archaeological site, a patent number and what does it protect?

Yes, 1,090,018 is a U.S. patent issued on March 10, 1914, to Joseph H. Bremer for "Pocket Ash-Tray."

This answer was found by looking the number up in the *Official Gazette* utility patent section. The same answer could have been found by looking up the number on the patent microfilm (or paper sets) in any PDL that has patents back this far.

3. What country issues patents with the SU prefix?

The Union of Soviet Socialist Republics. This answer must be found by using tools other than those from the Patent and Trademark Office. However, a number of common library tools will answer this basic question: *Chemical Abstracts, Index to Patents* (volumes contain a list of country prefixes), and the database user's guides for WPI (350, 351 on DIALOG) or INPADOC.

4. Did Mercedes-Benz receive a U.S. patent in 1951 for a safer automobile body frame?

Yes, U.S. patent 2,700,571 was issued to Bela Barenyi in 1955 for "Frame Structure of Automotive Vehicles" and assigned to Daimler-Benz. This patent claimed a priority application in Germany of February 17, 1951.

The important mental steps here were to remember or discover that Mercedes-Benz is the name of the car, not the name of the company, that this is a German car, and that the date given in the ad may have been when it was applied for, granted, etc. Once you had the name of the company, the next task was to look in the "Index to Patentees" of the *Official Gazette (OG)* for 1951, 1952, 1953, 1954, and 1955. This particular patent was indexed in the 1955 volume. The next step was to take the patent number found in the Index and look it up in the 1955 *OG* and read the entry. All data found in this entry correspond with what was gleaned from the television commercial. Therefore we can be fairly certain that this is what we are looking for.

CHAPTER 7

1. Is Jerome Lemelson a prolific inventor?

Yes, he has over 400 patents to his credit, including 4,169,303, "Fastening Materials" and 4,121,794, "Flying Toy." The answer was found by looking in the "Index to Assignees/Patentees." The answer could also have been found by doing a computer search on one of the commercial databases such as CLAIMS or WPI.

2. Does Buckminster Fuller have a patent on the geodesic dome?

Yes, it is 3,139,957, "Suspension Building." The problem here is that his first name is Richard. Remember that you must use your full legal name on the patent; not what you prefer to be called. Also note that the word "geodesic" is not used in the Claim.

The answer to this one is the same technique as problem 1 except that you will find no Buckminster Fuller. In checking his name in an encyclopedia or library card catalog, you will find that his first name is actually Richard.

3. Does the rock star Eddie Van Halen have a patent?

Yes, 4,656,917, "Musical Instrument Support." This answer is found using the same technique as discussed in number 1.

CHAPTER 9

1. Is a process for protecting videotapes from unauthorized copying protected?

Yes, two patents were granted to John O'Ryan, U.S. patent 4,631,603 for "Method and Apparatus for Processing a Video Signal so as to Prohibit the Making of Acceptable Video Tape Recordings" and U.S. patent 4,577,216, for "Method and Apparatus for Modifying the Color Burst to Prohibit Video Tape Recording."

2. Find if sunscreens for cars have been patented. If so, what class and subclass are they?

Abraham Levy has design patents 236,870, 236,869, 236,868, and 237,663, along with utility patent 4,202,396. What is interesting is that this seemingly simple invention has a great deal of patent protection. The utility patent number was found using the following steps, and the design numbers were retrieved from the utility number. While this example uses CASSIS CD-ROM, the same search could have been done using the *Manual of Classification* and a commercial database such as INPADOC, WPI, or CLAIMS.

Step 1: Think about how the invention of a car sunscreen works and list the important function or operative parts. They are:

a. folded material

b. held up by the visors

c. fits inside windshield

d. etc.

Step 2: Go to *Index to the Manual of Classification*. In the *Index* look up sunscreen, blinds, and other suggested terms.

Step 3: Go to the *Manual of Classification*.

a. Check class 160, "Closures, partitions and panels, flexible and portable" (see appendix 30).

b. Scan the schedule, reading the all caps first until reaching 84.1 "Pleating type."

c. Continue to the end of the schedule, selecting 130, 160, and digest 3.

Step 4: Go to the *Classification Definitions* and read the class description (see appendix 31).

a. Read class definition for 84.1, 130, and 160. Particularly try to find a definition of "lazy-tong" as this seems important to know.

b. Follow all cross-references.

c. Determine if these definitions are relevant—they are.

Step 5: Go to CASSIS CD-ROM (or commercial database) and input selected class/subclass combinations—in this case 160/84.1, 160/107, 160/130, and 160/136—and get a list of patents resident in each. An example of 160/107 is included in appendix 32, sample search for sunscreens.

Step 6: Scan printout and select a specific subclass to first look up the patent numbers in the *Official Gazette*. Pick the one that seemed most relevant from the *Manual*.

Step 7: Go to the *Official Gazette** and:

a. Read all abstracts or claims and characterize what is relevant or irrelevant about them.

b. Recheck the *Definitions* if necessary.

c. If you seem to be on the right track continue with Step 8.

Step 8: Check the full patent of those with relevant *OG* citations.

a. Read the patent abstract.

b. Take note of the "prior art."

c. Check field of search for other interesting subclasses.

d. Look carefully at the patent disclosure.

e. Read all the claims.

Step 9: Back to CASSIS CD-ROM with several patent numbers that seem relevant to get their original class/subclass. Thirty were selected and among them was 4,202,396.

Step 10: After much iteration: back to the *Manual*, the *OG*, the *Definitions*. And generally around and around examining nearly 150 *OG* entries and 17 complete patents. The answer was determined to be 4,202,396.

Step 11: The complete patent cites relevant design numbers. Check these and you have the complete answer. Another possibility is now that you have the inventor's name, do a commercial database search by inventor to retrieve other numbers relevant to the utility number just found.

---

*Since you are looking for an older patent, start examining *OG* entries from the bottom of the printout.

3. What pacemaker advances has Intermedics patented?

To find pacemaker advances patented by the company Intermedics since 1983 the searcher could look up Intermedics on the *Assignee/Patentee* microfiche and then examine each weekly issue of the *Official Gazette* for the time period not covered by the microfiche. This is the cheap answer, since it costs no money, only time. A faster answer would be to use any of the commercial databases and do an author (inventor) search.

4. Find the first patent for a satellite.

Following the same steps as used for answer 2, find that in 1958 U.S. patent 2,835,548 was granted to Robert C. Bauman for a "Satellite Structure."

5. Who invented earmuffs?

Chester Greenwood was granted U.S. patent 188,292 on March 13, 1877 for "Ear Protectors," later known as ear mufflers and then as ear muffs.

6. When were combination smoke and heat detectors invented?

U.S. patent 3,938,115 was granted in 1976 to Sidney Jacoby for "Combination Smoke and Heat Detector Alarm."

7. Was the computer punch card system patented?

Patent 395,782 was issued to H. Hollerith for "Art of Compiling Statistics." This was for punched cards, also known as Hollerith cards. This inventor had an interesting history—he invented these cards as a result of his job collecting data for the 1880 U.S. Census.

8. Has a machine been invented to accurately focus a beam of light on the wires of a computer chip?

Yes, patent 4,776,509, "Single Point Bonding Method and Apparatus," was granted to Daniel Andrews et. al of the Microelectronics and Computer Technology Corp. (MCC) on January 11, 1988.

CHAPTER 10

1. What is being protected by D280,670 and U.K. 1015921?

D280,670 was granted to Simon C. Fireman for "Pair of Cushioned Seals for Eyecups of Swimmers Goggles" in 1985. This was found by inputting this number in CASSIS. However, any commercial database or the *Official Gazette* would have given this information.

U.K. 1015921 was found by doing a search on DIALOG files 351 and 350. The only difficulty here is remembering that files 350 and 351 use the country code of "GB" rather than U.K. This number was issued to Benger Labs for "Benzamides."

Our conclusion is that the listed patent numbers protect different aspects of these swim goggles.

14 Bibliography of Patent Tools and Books

Research is to see what everybody else has seen, and to think what nobody else has thought.

—Albert Szent-Gyorgyi

This chapter consists of a list of patent reference tools that were discussed in chapters 6 and 8 and then were used in chapters 7 and 9. Care was taken in those chapters to describe several types of search strategies so that users can achieve "selective retrievability." However, because these tools are so important and complex the purpose of this chapter is to describe the tools one-by-one.

The bibliography is divided into two sections, first publications from the U.S. government on patents and then a short list of books on various topics of interest that relate to patent searching, such as how to write a claim, how to patent without an attorney, or how to prepare patent applications which are not published by the government.

GOVERNMENT PUBLICATIONS

Assignee/Patentee Microfiche. 1978-1987. Washington, D.C.: U.S. Patent and Trademark Office (PTO). Annual cumulation.

An alphabetical listing of all inventors and assignees at date of issuance since 1978.

Basic Facts about Patents. 1989. Washington, D.C.: Government Printing Office.

Free, brief, nontechnical information to help answer the most frequently asked questions about patents.

Catalog of Government Inventions for Licensing. 1987- . Springfield, Va.: National Technical Information Service. Annual. PB88-101951.

Consists of detailed summaries of all inventions announced in a particular year. These summaries are arranged into forty-four subject categories and include an inventor index.

Classification Change Orders. No date. Washington, D.C.: U.S. Patent and Trademark Office. Looseleaf. Updated as needed.

These consist of changes made to the *Manual of Classification, Class Definitions*, and patent numbers resident in a specific class and subclass. They are the result of reclassification projects and serve as interim guides to location until new definitions and schedules are issued.

Complete Patents. 1776- . Washington, D.C.: U.S. Patent and Trademark Office. $1.50/each for design and utility patents, $6.00/each for plant patents.

Complete or full patents are issued every Tuesday at the same time as the *Official Gazette*. Patents are also available from many commercial document delivery sources such as Airmail Pat, CTIC, Rapid Pat, etc.

Concordance—United States Patent Classification to International Patent Classification (IPC). 1990. Washington, D.C.: U.S. Patent and Trademark Office.

This tool can be used to get an idea of how U.S. classes relate to the IPC. International Patent Classification classes are printed on the first page of all modern U.S. patents and are often retrieved when searching the commercial databases. International Patent Classification is *not* given on CASSIS CD-ROM.

Definitions to the Manual of Classification. No date. Washington, D.C.: U.S. Patent and Trademark Office. microfiche. Updated as needed.

Provides detailed definitions of the subject matter to be found in or excluded from each class and official subclass. The definitions limit or expand in a precise manner the meaning intended for each class and subclass title.

Elias, Stephen R. 1989. *Patent, Copyright and Trademark Intellectual Property Law Dictionary*. Berkeley, Calif.: Nolo Press.

A dictionary of legal jargon and terminology associated with the law of trade secrets, copyright, trademarks, unfair competition, and patent procedures.

General Information Concerning Patents: A Brief Introduction to Patent Matters. 1989. Washington, D.C.: Government Printing Office.

Inexpensive introduction to patents, trademarks, and copyrights. Intended especially for inventors, prospective applicants, and students.

Index of Patents Issued from the United States Patent and Trademark Office. 1872- . Washington, D.C.: U.S. Patent and Trademark Office. Annual. 2 vs.

Volume 1 consists of an annual listing of inventions published that year by class and subclass. After a number of years the usefulness of this tool diminishes because of reclasses and changes to the schedules.

Volume 2 consists of an annual listing of inventors and assignees granted patents during the calendar year covered.

A companion to these is the *Index of Trademarks Registered with the United States Patent and Trademark Office*. It consists of an alphabetical listing of registrants.

Index to the U.S. Patent Classification. 1872- . Washington, D.C.: U.S. Patent and Trademark Office. Annual.

This is an alphabetical listing of subject headings referring to specific classes and subclasses. These headings are not an alphabetical inversion of those found in the *Manual*, but are instead a subjective list of relevant terms, phrases, synonyms, acronyms, and trademarks that have been selected over the years to aid the user in identifying and describing products, processes, and apparatus of patent disclosure. The *Index* is the entrance point for the *Manual*. It should be noticed that the *Manual* is updated as needed, but its *Index* is published annually. This can cause some dislocation that searchers should be aware of. Chapter 10 discusses ways to ameliorate this situation when it happens.

Manual of Classification. 1988- . Washington, D.C.: U.S. Patent and Trademark Office. Looseleaf. Updated as necessary.

Consists of looseleaf sections that list the numbers and descriptive titles of more than 400 classes and 100,000 subclasses used to classify or organize patented subject matter. This tool makes it possible to search for related subject matter by how something works.

Manual of Patent Examining Procedure (MPEP). 1989. Washington, D.C.: U.S. Patent and Trademark Office. 5th edition revised. 1987. Updated as necessary.

As the procedure manual for the Patent and Trademark Office, this includes related items from the U.S. Code, Code of Federal Regulations, and Patent Cooperation Treaty.

The office also issues a procedure manual for trademark examination called *Manual of Trademark Examination Procedures*.

The National Inventors Hall of Fame. 1985- . Washington, D.C.: U.S. Patent and Trademark Office. Annual.

Consists of biographies and drawings of U.S. patent holders whose inventions have contributed to the nation's welfare. As a result of this contribution each listee has been inducted into the National Inventors Hall of Fame.

Official Gazette of the United States Patent and Trademark Office (OG). 1872- . Washington, D.C.: U.S. Patent and Trademark Office. Weekly.

This is the official journal of the U.S. Patent and Trademark Office. The listings are in patent order number and include bibliographic information along with a drawing, if relevant, and either an abstract or the most comprehensive claim. The *OG* also contains inventor/assignee indexes, notices, listings of reissues and reexaminations, and the numbers of patents that have expired due to failure to pay maintenance fees.

The *OG* is issued in two parts each week. The second part is the official journal for trademarks and lists the trademarks published that week.

Patent Attorneys and Agents Registered to Practice before the U.S. Patent Office. 1987- . Washington, D.C.: U.S. Patent and Trademark Office. Annual.

Patent Information and Documentation in Western Europe; An Inventory of Services Available to the Public. 1989. 3rd rev. ed. New York: K. G. Saur.

Each country's entry includes general information on patent documents, official gazette, prices, register of legal status, and provincial libraries.

United States Code, 1989 Edition, Containing the General Permanent Laws of the U.S. 1989. Title 35. Washington, D.C.: Government Printing Office.

Title 35 of the U.S. Code is specifically concerned with patent law; Title 25, chapter 22, with trademarks, and Title 17 with copyrights.

United States Code of Federal Regulation. 1989. Washington, D.C.: Government Printing Office.

Title 37, chapter 1 deals exclusively with patents.

U.S. Serial Set. 1789-1969. Washington, D.C.: Government Printing Office.

Prior to 1843 there are no patent abstracts. Serial 207, Document 50, issued in 1830, contains a list of all U.S. patents granted through 1829. From 1843 on, numbers were assigned to patents beginning with number 2901. Access to those patents issued prior to the publication of the *OG* is through the *United States Government Publications; Serial Set* (also known as the *Congressional Series of United States Public Documents. Congressional Set*). *Checklist of United States Public Documents, 1789-1909* and the *CIS U.S.*

Serial Set Index. *CIS* indexes the American State Papers and the 15th-34th Congresses (1789-1857 and 1857-1879). These papers contain listings by number of U.S. patents issued during that session. See appendix 14 for an example.

The tools listed above can be found in any patent depository library (see appendix 13 for a list of all PDLs) and in many public and college or university libraries. The next tool can only be found in a PDL.

CASSIS CD-ROM. No date. Washington, D.C.: U.S. Patent and Trademark Office. Disks are updated quarterly.

Consists of all U.S. patent numbers issued since 1776, arranged by class and subclass. Abstracts are available for most patents issued in the last three years, and titles for most patents issued since 1976. However, it does not include the inventor's name. See appendix 33 for a list of U.S. patent numbers and their time frames.

NONGOVERNMENT SOURCES OF INFORMATION

This is a highly selective list of recent books.

Flanagan, John R. 1983. *How to Prepare Patent Applications*. Troy, Ohio: Patent Educational Publications. ISBN 0-913995-00-2.

Subtitled "A Self-study Course Book Using Actual Inventions."

Foltz, Ramon D. 1988. *Protecting Scientific Ideas and Inventions*. Boca Raton, Fla.: CRC Press. ISBN 0-944606-03-2.

Gausewitz, Richard L. 1983. *Patent Pending, Today's Inventors and Their Inventions*. Old Greenwich, Conn.: Delvin-Adair. ISBN 0-8159-6522-2.

Greene, Orville. 1979. *The Practical Inventor's Handbook*. New York: McGraw-Hill. ISBN 0-07-024320-4.

Covers protecting your invention, getting a patent, and how to get ideas "around the house," in "the kitchen" or "toys and games."

Greer, Thomas J. 1979. *Writing and Understanding U.S. Patent Claims*. New York: Michie Co. ISBN 0-87215-238-3.

This workbook presents a variety of examples of claims in the chemical, mechanical, and electrical fields.

Hale, Alan M. 1983. *Patenting Manual*. Buffalo, N.Y.: SPI Inc. ISBN 0-94341802X.

Written for independent inventors who want to learn about the processes of inventing and patenting.

Kivenson, Gilbert. 1982. *The Art and Science of Inventing*. 2nd ed. New York: Van Nostrand Reinhold. ISBN 0-442-24583-1.

Konold, William, et al. 1979. *What Every Engineer Should Know about Patents*. New York: Marcel Dekker. ISBN 0-8247-6805-1.

Levy, Richard C. 1990. *Inventing Patenting Sourcebook*. Detroit, Mich.: Gale Research. ISBN 0-8103-4871-3.

Subtitled "How to Sell and Protect Your Ideas."

Miller, Arthur R. 1983. *Intellectual Property, Patents, Trademarks, and Copyright*. St. Paul, Minn.: West Publishing. ISBN 0-314-74524-6.

Patents and Patenting for Engineers and Scientists. 1982. Piscataway, N.J.: IEEE Professional Communications Society. ISBN 0-87942-700-0.

Consists of a collection of articles reprinted from *IEEE Transactions on Professional Communications* PC-22, no.2 (1979).

Pressman, David. 1988. *Patent It Yourself*. 2nd ed. Berkeley, Calif.: Nolo Press. ISBN 0-87337-075-9.

Covers, among other topics, documentation, "how to write for Uncle Sam," and "now for the legalese—the claims."

Patents throughout the World. 2nd ed. 1978- . New York: Trade Activities. Looseleaf.

Consists of information by country on who may apply for a patent, when the application must be filed, membership in international conventions, requirements for applications, etc.

Rivkin, Bernard. 1986. *Patenting and Marketing Your Invention*. New York: Van Nostrand Reinhold. ISBN 0-442-27824-1.

Robinson, Judith S. 1988. *Tapping the Government Grapevine*. Phoenix, Ariz.: Oryx Press. ISBN 0-89774-179-X.

Chapter 7 covers patents, trademarks, and copyrights.

Samuels, Jeffrey M. 1989. *Patent, Trademark and Copyright Laws*. Washington, D.C.: Bureau of National Affairs. ISBN 0-87279-612-9.

A compilation of all U.S. laws that pertain to intellectual property.

Glossary

The vocabulary used in patent law and patent searching is very specific and sometimes arcane. As a help to readers the following glossary attempts to define the most frequently used and important words.

Abridgment: Synonym for abstract, used in particular in the United Kingdom.

Abstract: A concise summary of the technical disclosure of a patent document that enables a reader to quickly determine the subject matter covered.

Agent: A person who provides professional services in intellectual property matters. An attorney may act as an agent.

Applicant: The applicant is the person who files an application for the grant of a patent (or trademark).

Application, patent: A legal petition for a patent, describing an invention sufficiently so that an ordinarily skilled person could make or use it. Each application consists of a specification and one or more claims.

Art: A field or area of technology. See also **Prior Art**.

Art Group: An administrative arrangement of related United States classification numbers.

Assignee: The person or corporation to which patent rights are transferred, or assigned.

Assignment: A written transfer of property rights in an invention from the inventor (assignor) to someone else (assignee).

Assignor: The owner of the patent rights who transfers these rights to another by assignment.

Citation: A citation in a patent document is a reference to another document which may affect the novelty and/or obviousness of a (claimed) invention.

Claim: One or more statements at the end of a patent or application defining precisely the novel features of the protected invention.

Class Schedule: *See* **Schedules.**

Classification: In patent information "classification" means a specific system which subdivides technology into discrete units.

Composition of Matter: This statutory class of inventions includes chemical compounds and mixtures of single compounds.

Conception: The initial mental formulation of the elements of an invention.

Continuation: Refers to a patent application filed in the United States, where a second application may be made for the same invention as described in a prior application. The applicant and the disclosure presented in the continuation must be the same as in the original, which is subsequently abandoned.

Continuation-in-part: Refers to applications filed under the laws of the United States, where an application may be filed during the lifetime of an earlier application by the same applicant, covering a substantial portion or all of the subject matter of the earlier application and adding subject matter not in the earlier application.

Co-ordinate Subclasses: *See* the example listed under **Indent**. Co-ordinate are equal subclasses; they are not indents of each other. In the example, 2, 4, and 6 are co-ordinate.

Copyright: Every work of original authorship is protected by inherent copyright protection, which protects the mode of expression of an idea. It does not protect the idea itself. Copyright registration, which is optional, is in the Copyright Office of the Library of Congress. See also **Trademark**, and appendix 34 for a chart demonstrating some of the differences between copyright, trademark, and patents.

Cross-references: These are found in the *Classification Definitions* and serve as pointers to other relevant class/subclass combinations.

Cross Reference Art Collections: These are often listed at the end of specific schedules in the *Manual of Classification* and serve to group together patents that would otherwise be scattered throughout the file. Unlike *Digests*, these are all defined and inventoried.

Design Patent: Protects new, original, and ornamental designs for articles of manufacture.

Description of the Invention: Specifies the technical field to which the invention relates, includes a brief summary of the technical background of the invention, and describes the essential features of the invention with reference to the drawings.

Digests: Unofficial (undefined and uninventoried) groupings of patents that would otherwise be scattered throughout the schedules. These are the forerunners of **Cross Reference Art Collections** and as such are in the process of being discontinued.

Diligence: Demonstrated continuing activity following conception leading to reduction to practice of the invention, or to the filing of a patent application.

Disclosure: A description of the invention claimed in the patent document in a manner sufficiently clear and complete for the invention to be carried out by a person knowledgeable in the field ("person skilled in the art"). Sometimes call an enabling document.

Disclosure 2: This is an agreement drawn up by an attorney to protect the inventor's rights by proving his or her conception and reduction to practice prior to the submission of a patent application. There is no specific format requirement, but like the patent disclosure, it must be clear enough to allow a patent attorney to start preparing an application.

Drawings: The part of the disclosure that consists of pictorial representation of the various elements from top, bottom, and side of the invention.

Examination: Begins with the check of the application to make sure that it complies with the formal requirements (oath, description, drawing, claims, abstracts, fees) and whether the inventor's name, address, and nationality are present. In the second step, the examiner checks to see if the application complies with the patentability requirements of the law, and whether the claimed invention is new and unobvious.

Family: *See* **Patent Family**.

File (verb): To submit an application or amendments to the Patent and Trademark Office. Filing and application represent the first step in the procedure for obtaining a patent.

File-wrapper: The file maintained by the Patent and Trademark Office that contains all documents, proceedings, and correspondence related to the application. It is the totality of the written record and as such forms the history of examination.

Grant: The act of conferring a patent on an applicant after examination of the application.

Ideas: These are the tools of inventors, used in the development of inventions, not inventions per se. Consequently ideas are unpatentable—to be patentable ideas must be reduced to practice.

Indents: These are used in the schedules which make up the *Manual of Classification* to delineate how subclasses are arranged, for example:

(1) A
(2) with C
(3) and B
(4) with D
(5) and B
(6) with D

Items 2-6 are all indented under 1. It is also true that 3 is an indent of 2 but 4 is not.

Infringement (1): Direct infringement involves making, using, or selling within the term of the patent the entirety of the invention defined by a claim of the patent. Patent infringement is litigated through the federal district courts.

Infringement (2): Concerns using someone else's trademark without license or authorization, for example, using a trademarked cartoon figure as part of a sign for a daycare facility. Trademarks can be lost unless the owner practices diligence; they sometimes become descriptive nouns (e.g., dry ice, aspirin, shredded wheat, and cellophane).

Intellectual Property: The term denotes a wide variety of personal property rights, such as design, plant, and utility patents; trademarks; tradenames; fictitious names; copyrights; and trade secrets.

Interference: A Patent and Trademark Office procedure for determining which of two or more inventors should be entitled to pursue a patent. This procedure is used to determine, among the applicants, who was first to reduce the invention to practice.

International Patent Application: An application for the protection of an invention filed under the Patent Cooperation Treaty (PCT).

Invention: The conception of a novel and useful contribution followed by its reduction to practice. It may be a solution to a specific problem in a field of technology, it may relate to a product, it may be an apparatus, a process, or a new use. To be patentable it must be new and unobvious.

Inventive Step: A requirement of patentability meaning that the invention must be not only new but also unobvious in the sense that, having regard to the relevant prior art, it would not have been found by a person ordinarily skilled in the art.

Inventor: A person who devises or is the author of an invention.

Machine: One of the statutory classes of invention. A machine is an invention that has parts that move with respect to one another.

Manufacture, Article of: Statutory class of inventions in which the parts do not move with respect to one another.

New: *See* **Novel**.

Non-patent Literature: A phrase used for those scientific journals and periodicals that contain articles useful for patent searching and examination.

Novel (novelty): An invention that is not known or used by others, described in any printed publication, or anticipated by prior art.

Obviousness: This requirement for patenting maintains that if the invention was obvious at the time of invention to anyone ordinarily skilled in the art, then the invention is not patentable.

Opposition: Opposition is a request presented by the "opposing party" to the Patent and Trademark Office to refuse the application or annul the property rights.

Paris Convention: The Paris Convention for the Protection of Industrial Property is the most important international convention in the field of intellectual property, adhered to by over 100 countries. It provides, in part, that each contracting state must ensure that foreigners enjoy the same protection given to its own nationals and the right of priority established in Article 4 of the Convention. The Convention also lays down various minimum standards of protection and common rules which all contracting states must follow in their national law.

Patent: The name given to the document that grants legal protection to an invention, issued upon application and subject to meeting legal criteria. It creates a legal situation in which the patented invention can normally be exploited (manufactured, used, sold, imported) only with the authorization of the owner of the patent. In the United States protection is granted to design patents for fourteen years and utility and plant patents for seventeen years. The word *patent* is also used to describe the document (e.g., "I need a copy of patent 3,123,456.").

Patent Cooperation Treaty (PCT): An international treaty concluded in 1970 which provides for the possibility of filing international patent applications which have (in the signing countries) the same effect as a national patent application. Currently fourteen European countries are signatories. For details see WIPO official publication *PCT Applicant's Guide*, available from European Patent Office (Erhardtstrasse 27, D-8000 Muenchen 2, Germany).

Patent Family: Patent documents published in different countries but relating to the same invention are generally called a patent family. The members of a patent family describe the same invention but often in different languages.

Patentability: Not all inventions are patentable. Generally, national laws require that in order to be patentable the invention must be new, unobvious, and useful.

Plant Patent: Plant patents are issued to new varieties of asexually reproduced plants.

Preferred Embodiments: Another name for the disclosure section of the patent. In this section the reader finds the most information about how the invention works.

Prior Art: All relevant information including previously issued patents, articles, product catalogs, newspaper clippings, and books. Prior art consists of any subject matter bearing on the novelty of the claimed invention.

Priority: The first application filed by an applicant in any country has "priority filing." In some countries the first to file is granted a patent. In the United States the first to reduce to practice (i.e., conceive) is granted a patent. *See* **Interference**.

Process: The statutory class of patents that relates to an operation that may result in a composition of matter, machine, or article of manufacture.

Proximate Function: The concept that groups together inventions that work alike. Proximate is used in the context of nearest, next to, or next after.

Reduction to Practice, Actual: The act of completing an invention and testing it as needed to assure that it is operable under normal conditions. Actual reduction to practice is not a requirement for filing a patent application.

Reduction to Practice, Conception: The act of filing a patent application that includes a complete description of the invention.

Reissue Patent: A patent issued to correct and supersede a previous patent. It expires on the same date that the previous patent would have expired.

Search: An investigation of the applicable prior art in the field of invention to determine if the idea is new or has been anticipated.

Schedules: A shorthand way of referring to the classification schedules found in the *Manual of Classification*. Each schedule reflects a specific art and is arranged in hierarchical order, beginning with the most complex inventions at the top and proceeding to the least complex (i.e., having the fewest parts).

Skilled in the Art: A hypothetical individual familiar with the technology in his or her field is "one skilled in the art."

Specification: The part of a patent document which gives a detailed description of the invention and is accompanied by claims. It should include title, a brief summary of the invention, a description of the drawings, and a detailed description of the invention.

Statutory Bars: Activities specified by law that bar the granting of a patent. The most common are publication, sale, or use of the invention more than twelve months prior to application.

Statutory Invention Registrations: A defensive class of documents which covers inventions that could have been granted patent protection. However, the inventor chose to "disclose" the invention and not seek protection, thereby precluding others from getting a patent on this idea.

Trademark: The term is often used collectively to refer to commercial, service, or certification marks, covers any word, name, symbol, sound, device, configuration, or combination of these that is used to identify and distinguish goods and services from each other. Trademark rights may be used to prevent others from using a confusingly similar mark but not to prevent others from making the same goods or from selling them under a nonconfusing mark. Trademark registrations give very different protection from that granted by patents. Patent protection is limited, but rights in a trademark can continue indefinitely as long as fees are paid and diligence is applied to protection of the mark. Trademark registration is possible within the Patent and Trademark Office and in some state trademark offices. See appendix 34 for a chart demonstrating some of the differences between patents, trademarks, and copyrights.

Useful: To be useful an invention must be capable of achieving some minimal useful purpose that is not mischievous, immoral, frivolous, or injurious to the well-being, good policy, or sound mores of society.

Utility Patent: Utility patents grant a limited governmental monopoly to new, useful, and unobvious processes, machines, compositions of matter, and articles of manufacture.

Validity: Validity of an issued patent.

X-art Collections: *See* **Cross Reference Art Collections**.

Appendix 1
Sample Utility Patent

United States Patent [19]
Gault et al.

[11] **Patent Number: 4,517,465**

[45] **Date of Patent: May 14, 1985**

[54] **ION IMPLANTATION CONTROL SYSTEM**

[75] Inventors: **Roger B. Gault; Larry L. Keutzer,** both of Austin, Tex.

[73] Assignee: **VEECO/ai, Inc.,** Austin, Tex.

[21] Appl. No.: **480,095**

[22] Filed: **Mar. 29, 1983**

[51] **Int. Cl.**³ **A61K 27/02;** H01J 37/00

[52] **U.S. Cl.** **250/492.2**

[58] **Field of Search** .... 250/492.2, 398, 397, 250/442.1; 219/121 EB, 121 EM

[56] **References Cited**

U.S. PATENT DOCUMENTS

3,689,766	9/1972	Freeman	250/492.2
3,778,626	12/1973	Robertson	250/492.2
4,011,449	3/1977	Ko et al.	250/492.2
4,021,675	5/1977	Shifrin	250/492.2
4,234,797	11/1980	Ryding	250/492.2
4,383,178	5/1983	Shibata et al.	250/442.1

Primary Examiner—Bruce C. Anderson

Attorney, Agent, or Firm—Morgan, Finnegan, Pine, Foley & Lee

[57] **ABSTRACT**

A control system is disclosed for an ion implantation system of the type in which the wafers to be implanted are mounted around the periphery of a disk which rotates and also moves in a radial direction relative to an ion beam to expose successive sections of each wafer to the radiation. The control system senses beam current which passes through one or more apertures in the disk and is collected by a Faraday cup. This current is integrated to obtain a measure of charge which is compared with a calculated value based upon the desired ion dosage and other parameters. The resultant controls the number of incremental steps the rotating disk moves radially to expose the adjacent sections of each wafer. This process is continued usually with two or more traverses until the entire surface of each wafer has been implanted with the proper ion dosage.

9 Claims, 2 Drawing Figures

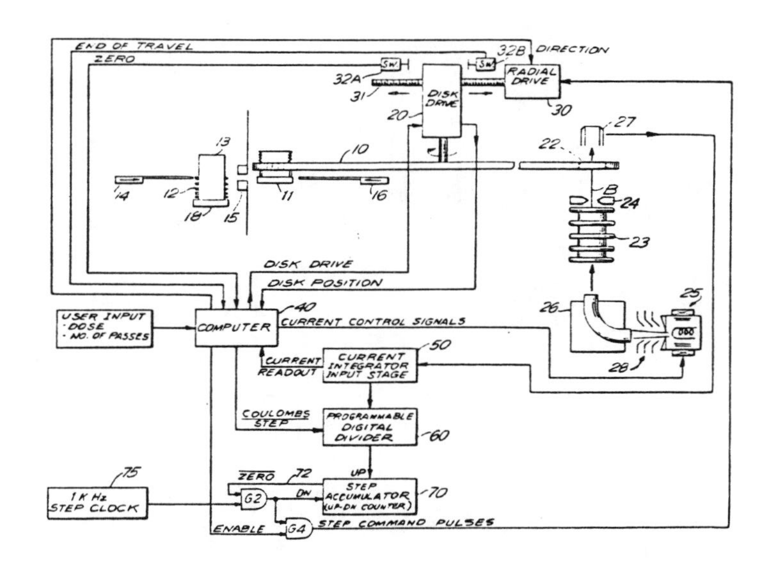

U.S. Patent May 14, 1985 Sheet 1 of 2 **4,517,465**

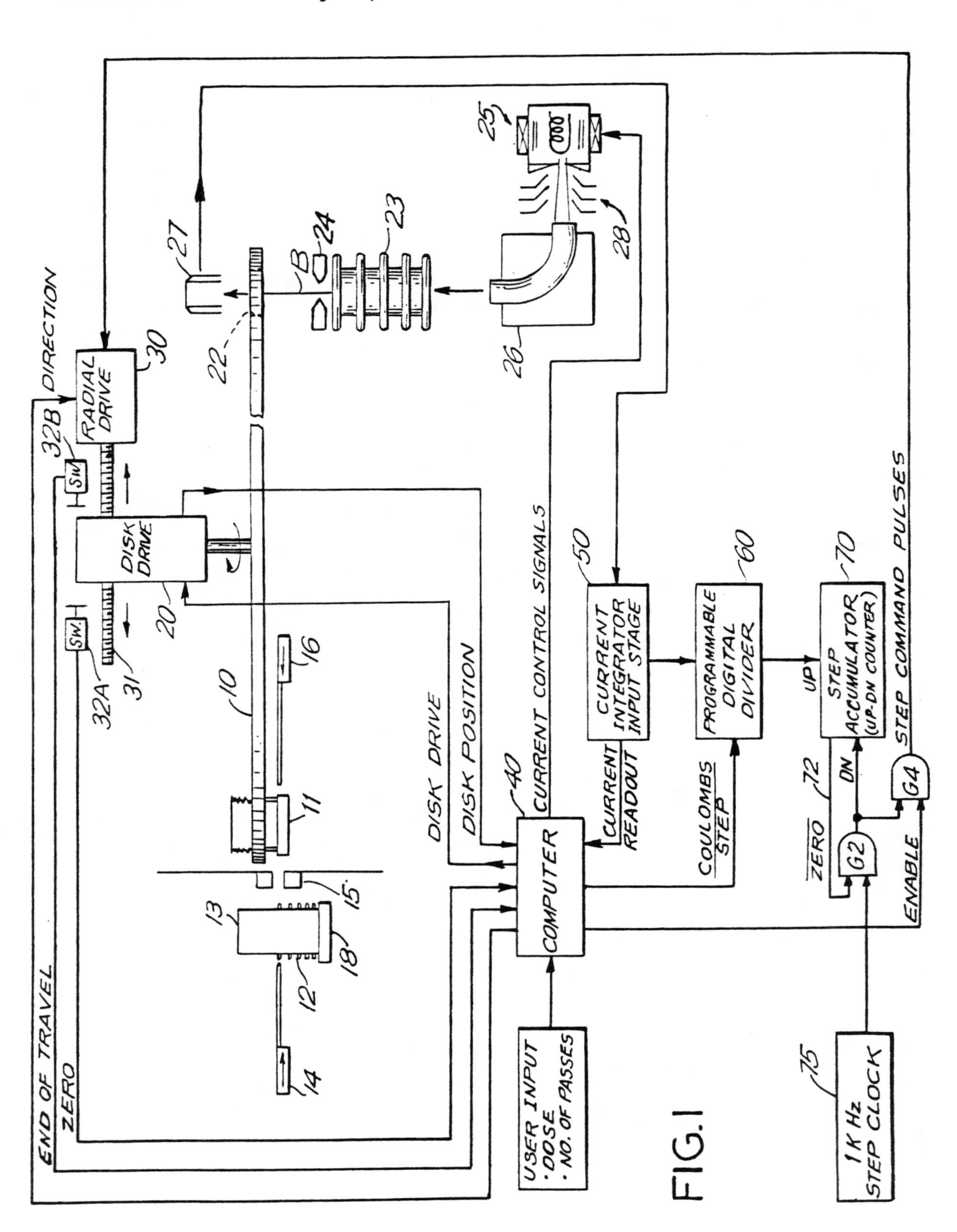

U.S. Patent May 14, 1985 Sheet 2 of 2 4,517,465

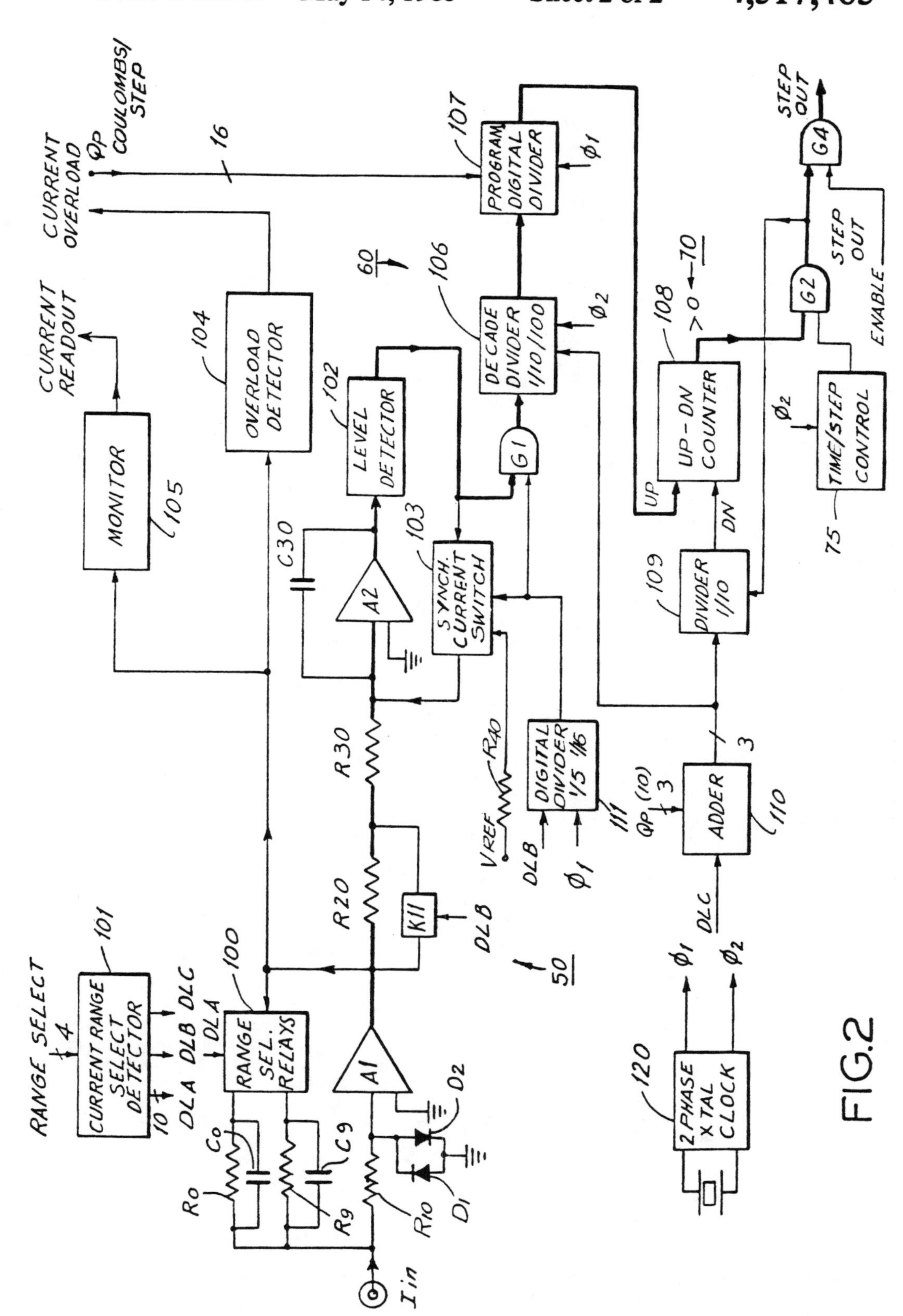

FIG.2

1

ION IMPLANTATION CONTROL SYSTEM

BACKGROUND OF THE INVENTION

1. Field of the Invention

The invention involves improvements in ion implantation techniques. It is directed to systems for controlling the ion dosage of work pieces such as semiconductor wafers. These systems utilize a rotary disk which carries the wafers to be irradiated around its periphery and rotates them past an ion beam. As the wafers are irradiated, a second mechanical displacement of the disk occurs in a radial direction so that the entire surface of each wafer is subjected to the beam and receives the proper dosage.

2. Description of the Prior Art

A known technique for implanting ions in wafers utilizes a rotating disk mounted in a vacuum chamber and adapted to carry a set of wafers around the periphery of the disk such that by rotating the disk a portion of each wafer is successively exposed to an ion beam which is projected into the chamber. A second drive system moves the rotating disk assembly in a radial direction so that successive adjacent tracks of each wafer are subjected to the ion beam until the entire surface of each wafer has been implanted.

Patents illustrating this technique include U.S. Pat. No. 4,234,797 issued to Jeffrey Ryding and U.S. Pat. No. 3,778,626 issued to Gordon Ian Robertson.

In one mode of the control system of the Robertson patent, the ion beam is monitored by measuring the current which flows to the disk itself by reason of its interception of the beam. This current is measured through a slip ring system and is supplied to a control system which also senses the radial displacement of the disk relative to the beam. Also supplied as an input to the control system is a desired dosage signal. With these inputs the Robertson control system controls the radial motion of the disk to control the ion distribution over the wafer surface.

The Ryding patent employs a current measurement arrangement which uses, instead of slip rings, a Faraday cage mounted behind the disk and located to receive pulses of the beam current when an aperture in the disk passes in front of the cage exposing it to the ion beam. Ryding employs the measured beam current to control the radial velocity of the rotating disk for the purposes of achieving a uniform ion dosage. While Ryding's technique eliminates the need to sense radial displacement, his process depends upon maintaining angular velocity at a constant value thus introducing a possible source of error.

SUMMARY OF THE INVENTION

The applicants have discovered that controlled uniform ion dosages can be achieved without the need for measuring or sensing radial displacement and without the need to depend upon a constant angular scan velocity. Rather than controlling radial velocity, the control system disclosed herein produces a a number of fixed incremental displacements in a radial direction in dependence on the integration of sensed beam current relative to a desired dosage level. This technique thus does not depend on a constant angular velocity nor does it require sensing of radial displacement.

2

PREFERRED EMBODIMENT

Serving to illustrate a preferred embodiment of the invention are the drawings of which:

FIG. 1 is a schematic diagram of the system illustrating its general geometry as well as the components of the control system and

FIG. 2 is a schematic diagram of the radial motor control circuits.

As shown in FIG. 1, a rotatable disk 10 mounted inside a vacuum chamber, not shown, includes a series of wafer clamp assemblies 11 for receiving wafers 12 from a wafer carrier 13.

During the loading cycle of the system, the disk 20 is rotated through a series of successive positions to align successive wafer clamps with the wafer carrier 13. As each position is reached, a loading transfer actuator 14 transfers a wafer 12 from carrier 13 through a wafer valve port 15 to the aligned wafer clamp 11 the underside of which is open to expose the major part of the wafer mounted in it. As each position is loaded, an elevator assembly 18 associated with the wafer carrier elevates the next wafer into alignment with the loader 14. The process continues until all the wafer sites are loaded.

After implantation, a generally similar unloading process occurs. The disk 10 is positioned to successively align each wafer site with the wafer valve and with the wafer carrier now arranged and progressively elevated to receive the implanted wafers through the action of unloading transfer assembly 16.

For achieving the foregoing loading and unloading movements of the disk, a position servo arrangement is employed with a resolver located in disk drive 20 serving as the disk position feedback sensor whose output is used to null out the digital command signal, supplied by computer 40, when the disk is in its commanded position. In the illustrative embodiment there are provisions for 15, 20, 25 or 35 wafer positions.

For implanting the wafers, an ion source 25 is utilized. The generated ions are accelerated out of the source with the aid of extraction electrodes 28 and directed to an analysis magnet 26 which subjects the beam to a 90 degree turn. It thereafter passes through an accelerator tube 23 to the analysis slit 24. The resultant beam B radiates the wafers on disk 10 as the latter rotates (at 500 RPM in the preferred embodiment). The beam is generally rectangular in cross section, its long dimension being wider than slot width and being oriented tangentially to the disk and wafers. Its short dimension is a fraction (e.g. $\frac{1}{4}$) of its length and of slot length (e.g. 1/20). An illustrative slot configuration is 9 mm×125 mm, with the long dimension oriented radially and being larger than the wafer diameter.

For sensing beam current, disk 10 is equipped with one or a number of slots 22 which exposes a current collector 27 to beam B during each revolution of disk 10. The measured current is used to control radial movement of the disk 10 and its wafers relative to beam B. The collector is sized to respond to the ion beam radiation over the entire slot length.

This radial motion is accomplished with a radial drive assembly 30 having an output worm shaft 31. Disk drive 20 is coupled to the worm and is displaced in a stepwise fashion by the radial drive which is controlled to function as a stepper motor. The assembly also includes limit detecting switches 32A and 32B which signal the computer when the extremes of a traverse are reached.

The step movement of the radial drive is controlled by the Integrator, Divider and Accumulator stages of FIGS. **1** and **2** while its direction is controlled by computer **40** which counts the passes and keeps track of translation and direction of traversal. (One pass is defined as a traversal from the zero displacement position to full displacement and back again).

In the illustrated embodiment each incremental movement of the radial drive produces a disk displacement of 1/1000th of an inch The mode of operation of the control system is to measure dosage as a function of the number of ions irradiating the slots **22** and to cause this incremental radial movement of the disk as a function of that dosage relative to a predetermined level.

To implement this technique, the system receives as one input the dosage level desired by the user. The latter may also specify the desired number of passes. If defaulted, the computer calculates a minimal integral value for this parameter.

Utilizing these inputs (dosage being in ions/cm^2) computer **40** computes the required accumulated charge developed in the slot area for each radial increment. This calculation, expressed in terms of coulombs per step, is given by the equation:

$$\frac{\text{Coulombs}}{\text{Step}} = \frac{K \cdot \text{Dose} \cdot Ai}{2 \cdot N} \quad (1)$$

where K is the charge constant 1.6×10^{-19}; D is dose in ions/cm^2, Ai is the sensed area, being the product of total slot width and step size, and N = the number of passes.

In addition to the charge/step computation, computer **40** performs the beam current calculation and, as shown in FIG. **1**, supplies the appropriate current control signal to ion source **25**. This source is preferably of the type in which the desired ions are derived from a plasma discharge process. An appropriate beam current to achieve the desired result is calculated such that the radial motor stepping pulses occur at an average rate of 500 per second when the disk is in the radial drive zero position. Thus:

$$I_{req} = 500 \cdot \frac{\text{Coulombs}}{\text{Step}}$$

Initiation of the operational phase includes rotation of disk **10** as a consequence of disk drive signals sent to disk drive **20** from computer **40**. When rotation has been established, beam current is sampled at each slot once per revolution, e.g., by collector **27**, and fed to current integrator input stage **50**. The current integrator generates in turn a train of pulses, the number of which is proportional to the integral of the current pulse sensed by collector **27**.

This train of pulses is fed to the programmable digital divider **60** which also receives from computer **40** the previously calculated value of required coulombs per step. The divider combines these inputs to derive the appropriate step pulse count, representing the increments the disk should be moved based upon the amount of charge which passed through the slots **22**.

This count is stored in a Step Accumulator **70** which is an up-down counter in the illustrated embodiment.

A Step Clock **75** operating for example at a 1 KHz rate, is used to remove (count down) steps from the Accumulator. Any time there are accumulated steps, the Step Clock is gated with line **72** in gate G2 to decrement the count at a 1 KHz rate. If the computer **40** has enabled the drive, these Step Clock pulses are gated through gate G4 to the Radial Drive **30** as Step Command pulses.

Since the computer set up the system to produce an average pulse rate of 500 pulses per second, the Step Accumulator **70** is emptied in approximately ½ the time between successive slot passages. As a consequence, the radial scan motion is discontinuous. It consists of alternating segments of constant radial velocity (e.g., 1 inch/second) and zero radial velocity. In terms of displacement an intermixing of spiral and constant radius segments' is produced.

This stepwise operation typically continues for a full traverse of the disk drive from zero displacement to full displacement and back again to the zero position.

As the distance between the disk axis and the beam decreases the tangential velocity of the active slot section decreases. Since the width of the slots is constant, the time during which the beam passes through them increases, and the integral of the current pulse increases proportionally. This, in turn, increases the number of pulses sent to the Step Accumulator. Since the Step Clock frequency is constant, the time required to empty the Accumulator increases and the section of the scan motion in which the velocity is 1 inch/second increases to compensate for the change in tangential velocity.

When the correct dosage has been applied to the entire surface of the wafers on disk **10**, scanning ceases and the wafers are unloaded through valve **15** to the carrier **13** as previously described.

Further details of the Current Integrator, Divider and Accumulator circuits are provided in FIG. **2**. As seen therein, the Integrator includes an input amplifier **A1** which receives collector cup current, I_{in} via an input resistor **R10**.

Any one of ten current measuring ranges can be selected in the circuit of FIG. **2** under control of a Range Select input to a Current Range Select Decoder **101**. The latter provides three outputs. DLA provides drive to a set of range select relays **100** which selectively switch an appropriate feedback network **R0**, **C0**; **R1**; **C1**; . . . **R9**, **C9** between the output and input of amplifier **A1** to provide the proper gain and frequency response for the selected range.

Range is selected by the computer **40** which receives a signal from an Overload Detector **104** driven by the output of amplifier **A1**. The computer is programmed to vary the range selection in the direction of increasing sensitivity until an overload is detected. When the overload is signalled, the computer backs down two steps by appropriate switching of the Range Select input thereby setting the input amplifier at its maximum sensitivity for safe linear operation.

Also driven by input amplifier **A1** is a monitor circuit **105** which supplies a beam current readout signal.

The output of amplifier **A1** is fed to the integrating stage **A2** via resistors **R20** and **R30**. The former is switched out of the circuit for some range values through a relay **K11** operated by the DLB output of decoder **101**.

The integrating stage includes an operational amplifier **A2** with its feedback integrating capacitor **C30**. The output of A2 is supplied to a level detector **102** which in addition to supplying an output to gate **G1**, also drives a synchronous current switch **103** whose output is fed back to the input of amplifier **A2**. The current switch **103** also receives a reference current input via a resistor

R40 energized from the V_{ref} supply. The switch in essence discharges the integrated charge in steps such that the number of steps and their resultant pulses, represent the integral of the input.

The current switch is also controlled as a function of range by a digital divider **111**. The divider receives clock input Ø1 along with an input DLB from the Range Decoder **101** which represents range coefficients of either three or ten. Divider **111** provides division by either 5 or 16 depending upon the selected range and also supplies the second input to AND gate G1.

The pulses produced at the other input to G1, representing the integral of the sampled current supplied at I_{in}, when ANDed are fed to the Programmable Digital Divider **107** via a Decade Divider **106**. The latter is controlled by an input received from an adder **110** which receives one input DLC from the range decoder **101** and a second input from one line of the Q_p output (coulombs per step) of the computer **40**. The DLC input to adder **110** is set for range values of 10^{-3} to 10^{-7}. The output of Adder **110** determines whether Decade Divider **106** divides by one, by ten or by one hundred.

Decade divider **106** is clocked by the Ø2 clock signal and supplies its output to the Programmable Digital Divider **107** which is implemented with a counter and also receives the computed coulombs-per-step value, Qp. It is strobed by the Ø1 output of the clock and its output constitutes count UP signals for the Up/Down Counter **108**. As previously noted, the output of counter **108** is ANDed with step clock **75** in gate G2. The output of G2 is ANDed in turn in gate G4 with the ENABLE line of computer **40** to provide the step drive to the radial motor.

The output of G2 is also fed back to a divider **109** which receives a range dependent input from adder **110** and supplies the down DN terminal of counter **108** for the purposes previously described.

By way of illustrating the operation of the system consider the case where disk **10** has 5 slots and rotates at 500 RPM. Assume the desired dosage is 2×10^{15} ions/cm^2, the selected (or computed) number of passes is 13, and the incremental slot area is 1.13×10^{-2} cm^2. As a result the charge per step is 1.390×10^{-7} Coulombs/Step.

Assume a particular beam current condition such that the integrator input is a train of 5 ma pulses. Under these conditions, a representative output from the integrator would contain 300 pulses. The programmable divider would be set to divide this output by 25, producing a count **12** in the accumulator. This in turn would cause 12 step pulses to be sent to the radial drive to displace the disk 0.012 inches.

What is claimed is:

1. In an ion implantation system having an ion beam, a rotating wafer support and displacement means for displacing the rotating support in a radial direction to expose successive sections of wafers carried on said support to said ion beam as the rotating support is displaced radially, the improvement comprising:

(a) means for computing a signal indicative of charge delivered to a section of said support;

(b) means for generating a signal indicative of desired dosage;

(c) displacement control means responsive to said charge and dosage signals for producing a signal for causing a sequence of predetermined incremental displacements of said support by said displacement means, and wherein each incremental displacement occurs at substantially the same velocity with the number of said incremental displacements being responsive to the values of said charge and dosage signals.

2. The apparatus of claim **1** in which said section of said support comprises an area defined by a slot in said support and the dimension of said predetermined increment.

3. The apparatus of claim **1** in which said incremental radial displacements are produced repetitively in reciprocal directions until a total desired dosage is obtained.

4. The apparatus of claim **1** in which said support includes a slot irradiated by said beam and in which the dosage signal represents desired dosage, slot width and increment length.

5. An ion implantation system having an ion beam, a rotating wafer support having an aperture, and means for rotating the support to expose successive wafers carried on said support to said ion base, the improvement comprising:

(a) radial displacement means coupled to said wafer support and configured to produce substantially constant velocity stepwise radial displacements of the support to expose successive sections of said wafers to said beam;

(b) measuring means positioned to be periodically radiated by said beam via said aperture for computing a signal indicative of charge delivered to a section of said support;

(c) means for generating a signal indicative of desired dosage;

(d) means responsive to said desired dosage signal for computing a signal representing desired dosage per step;

(e) displacement control means responsive to said charge and dosage per step signals for computing the required number of step drive pulses for said radial displacement means whereby said support undergoes a computed number of substantially constant velocity increments.

6. The system of claim **5** in which said measuring means produce a pulse train having a number of pulses related to said delivered charge.

7. The system of claim **6** in which said displacement control means include a digital divider for dividing said charge signal by said dosage per step signal.

8. The system of claim **7** including clocked gating means for gating the output of said divider to said radial displacement means.

9. The system of claims **5**, **6**, **7** or **8** including means for adjusting the intensity of said ion beam.

* * * * *

Appendix 2
Sample Design Patent

UNITED STATES PATENT [19]
Ng et al.

[11] **Patent Number: Des. 308,922**
[45] **Date of Patent: **Jul. 3, 1990**

[54] ADJUSTABLE PRINTER STAND

[76] Inventors: Michael C. M. Ng; Kheng L. Tan, both of Block 29, #01-132 Defu Lane 10, Singapore, Singapore, 1953

[**] Term: 14 Years

[21] Appl. No.: 928,385

[22] Filed: Nov. 10, 1986

[52] U.S. Cl. D6/429

[58] Field of Search D6/429, 430, 431, 425, D6/480, 486, 474; 108/92

[56] References Cited

PUBLICATIONS

Structural Concepts Corp., Jan. 1, 1981, p. 3, Printer Stand, left center.

Equipto Electronic Corp., inside front cover, Printer Stand, top left.

Primary Examiner—Bruce W. Dunkins
Assistant Examiner—Ann Hunt
Attorney, Agent, or Firm—Finnegan, Henderson, Farabow, Garrett & Dunner

[57] CLAIM

The ornamental design for an adjustable printer stand, as shown and described.

DESCRIPTION

FIG. 1 is a front perspective view of the adjustable printer stand showing our new design;
FIG. 2 is a left side view, the right side being a mirror image thereof; and
FIG. 3 is a front perspective view thereof.

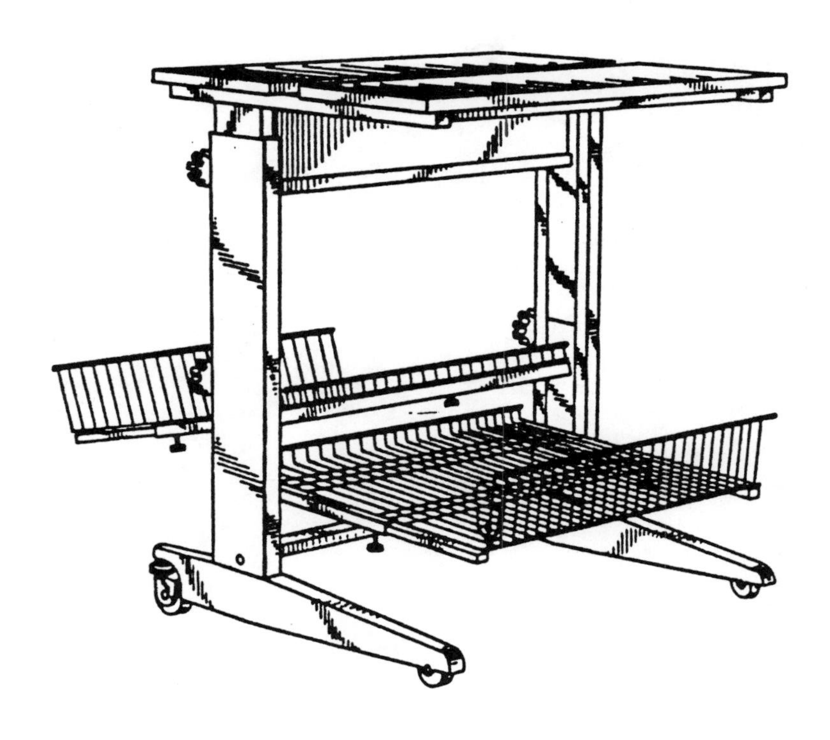

FIG. 1

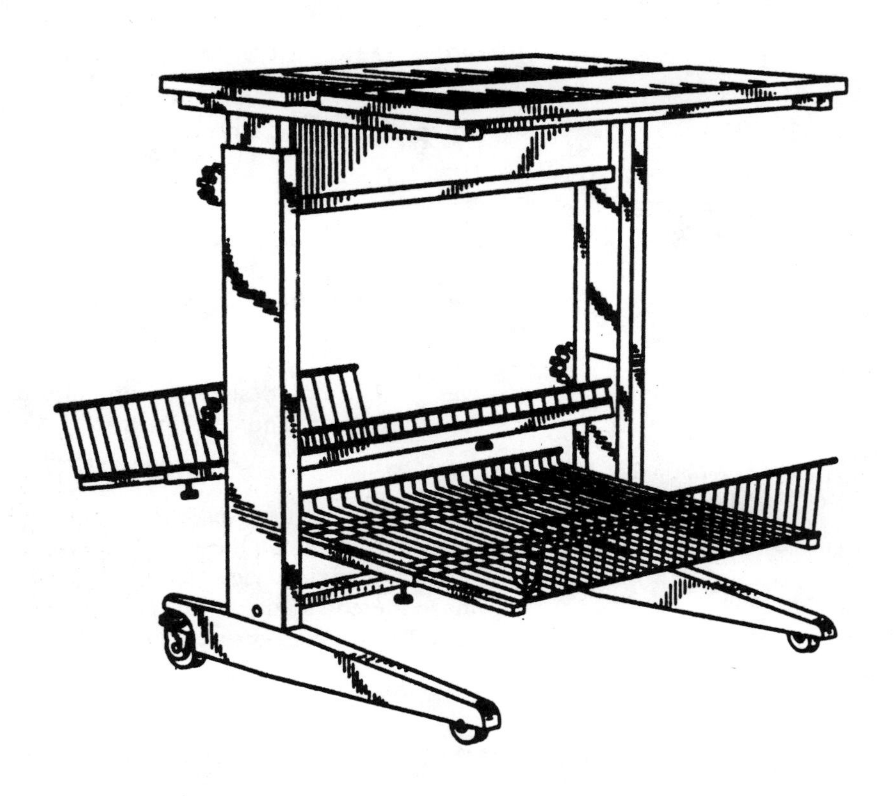

FIG. 2

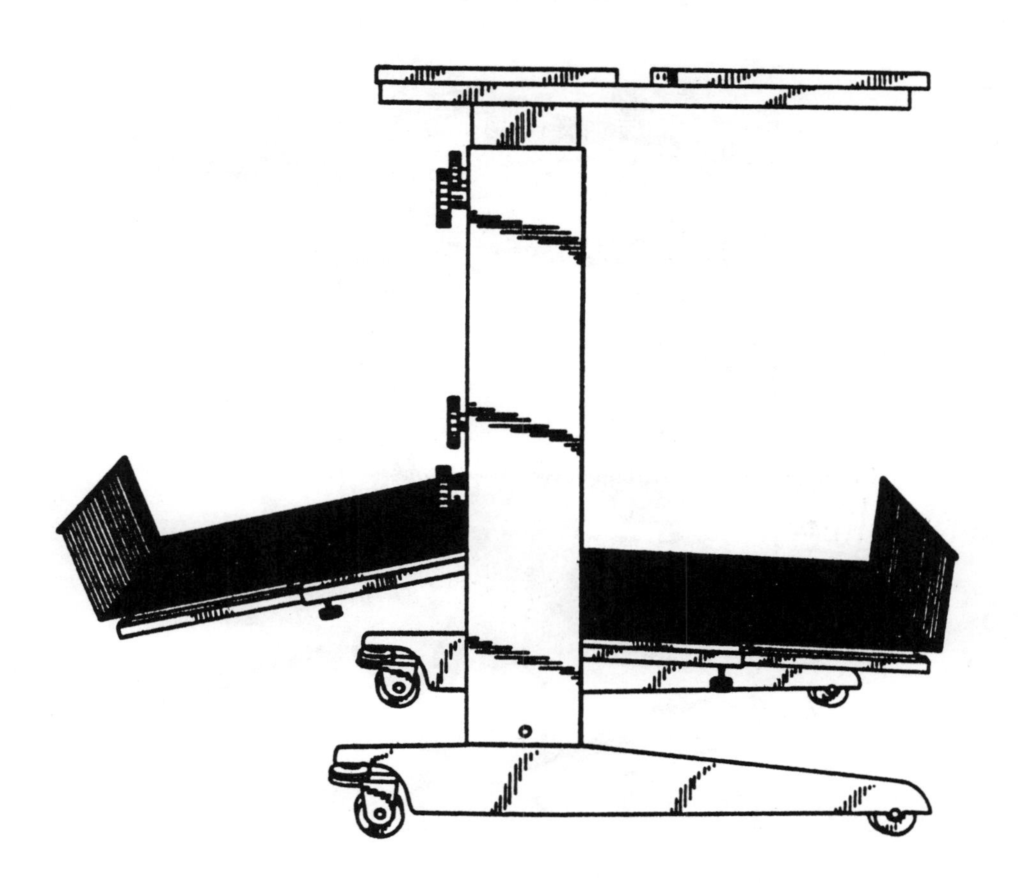

U.S. Patent **Jul. 3, 1990** **Sheet 3 of 3** **D308,922**

FIG. 3

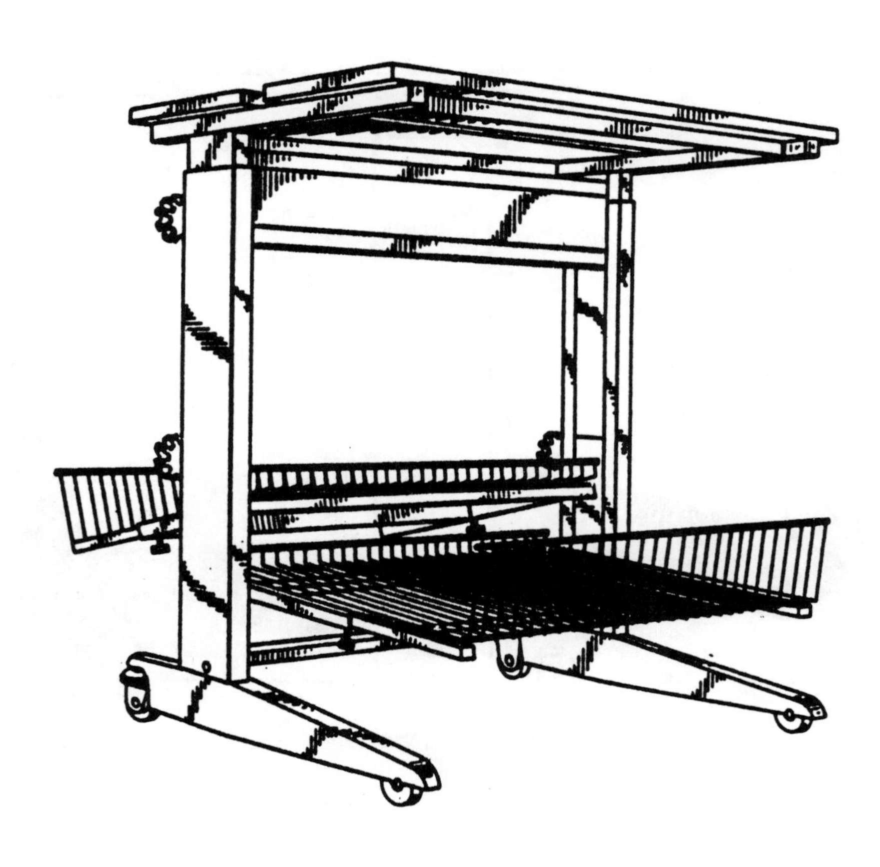

Appendix 3
Sample Plant Patent

United States Patent Office

Plant Pat. 3,045
Patented Apr. 13, 1971

1

3,045
APPLE TREE

Ralph Banta, Green Forest, Ark., assignor to Merle J. Lucas, doing business as Green Forest Nursery, Green Forest, Ark.

Filed June 18, 1969, Ser. No. 834,564
Int. Cl. A01b *5/03*
U.S. Cl. Plt.—34 **1 Claim**

ABSTRACT OF THE DISCLOSURE

This disclosure concerns a new and distinct variety of apple tree characterized by the fruit having a dark red uniform color and having an elongated shape with heavier stems than the Jonathan tree.

BACKGROUND OF INVENTION

This new and distinct variety was discovered by me on my orchard in Green Forest, Ark. (Route 1). About 50 to 60 cuttings have been taken from the parent tree and propagated and interplanted with various other varieties of apple trees, and are producing abundantly, true to form every year. The apples ripen two weeks to a month ahead of Jonathan apples and are redder in color and larger, and having a tartness, slightly similar to a Jonathan apple. Each year, since 1950, the tree has had the same growing habits and same fruit.

DESCRIPTION OF DRAWING

The accompanying drawings, in full color, show typical examples of fruit and foliage of my new variety of apple tree, the fruit being shown in axial and transverse cross section as well as in elevation.

DESCRIPTION OF THE NEW PLANT VARIETY

The following is a detailed description of my new variety with the color designations according to Maerz and Paul's "A Dictionary of Color."

Origin: Seedling.
Parentage: Unknown.
Classification: Hybrid.
Growth habit: Vigorous, upright, hardy and very fast growth habit during the first 5 years, at which time the growth slows down.
Height: Semi Dwarf, 18 to 20 feet.
Trunk: Thick and heavy, from ground up about 10 inches.
Branches: Very heavy and vigorous extending outwardly at an angle of 50 to 60 degrees.
Leaves: Abundant and large.
Size: Length—4 to 6"; width—1½ to 2½".
Shape: Ovate.
Margin type: Deltoid, serrated.
Color: Upperside—Plate 12, L–4; underside—Plate 21, G–4.

2

Lenticels: Large and abundant.
Flowers: About mid-April in northern Arkansas with larger blossoms than the Jonathan apple tree, having creamy pink color.
The fruit:
Shape.—Oblong ovoid, generally symmetrical.
Size.—About 3 inches diameter and about 3 inches high.
Color: Overall dark red, Plate 4, L–6 with relatively small spots Plate 9, 11 appearing sporadically on surface. Area around stem base, Plate 18, J–2.
Stem: About ¾ inch long.
Color: Plate 7, E–10 with overtones of Plate 8, H–7. Base of stem, Plate 13, L–1.
Cavity: About ½ inch deep and approximately 1¼ inches wide.
Basin: About ⅜ inch deep and about ⅜ inch wide.
Core: Seed cavity.—About 1 to 1⅛ inches wide and about 1 inch long. Color: Plate 9, F–1. Seeds—About ⅜ inch long and ⅛ inch wide and thick. Color Plate 12, L–10 and shaded outwardly to Plate 12, C–11 and Plate 8, C–6.
Flesh: Firm in texture, juicy and tart. Crunchy when eaten.
Aroma: Very pleasant and predominant.
Color: Slightly off-white.
Maturity season: In northern Arkansas from August 1st to 10th.
Taste: Tart, acid, tangy, somewhat sweet.
Keeping quality: Fair.
Use: Excellent eating apple. Very good for juice, cooking and canning. Apple when cooked stays in piece form and not reduced like apple sauce.

In particular my new variety of apple is distinguished by its deep overall red color. It ripens approximately two weeks before the Jonathan apple, has fair keeping quality and is particularly good for both eating and cooking purposes because of its pleasant taste and aroma which persists through the cooking process. The plant is distinguished by its rapid growth and by its high production of fruit of a size larger than the Jonathan. The branches grow from 18 to 24 inches of new wood each year. The variety is resistant to scab and blotch. Having thus described my new variety of apple tree, I claim:

1. A new and distinct variety of apple tree as shown and described, particularly characterized as to novelty by the large size, deep red color with the tart taste of Jonathan apple, by the early maturity season for its fruit, by the abundance of fruit each year and by the persistent pleasing aroma of the fresh fruit and throughout cooking and canning process.

No references cited.

ROBERT E. BAGWILL, Primary Examiner

April 13, 1971 R. BANTA **Plant Pat. 3,045**

APPLE TREE

Filed June 18, 1969

INVENTOR:
RALPH BANTA

WITNESS: Ruth Ingeborg Andris

BY: Cummins and Snow
ATTYS.

Appendix 4
Patent Fees as of November 5, 1990

PATENT FEES
35 U.S.C. § 41(a) and § 41(b)
Effective November 5, 1990

FEE CODE LG / SM	37 CFR	DESCRIPTION	LARGE ENTITY FEE INCL 69% SURCHARGE	LARGE ENTITY OLD FEE	SMALL ENTITY FEE INCL 69% SURCHARGE	SMALL ENTITY OLD FEE
101 / 201	1.16(a)	Basic fee for filing an application for an original patent, except design or plant case	$630	$370	$315	$185
102 / 202	1.16(b)	In addition to the basic filing fee, for filing or later presentation of each indep. claim in excess of 3	$60	$36	$30	$18
103 / 203	1.16(c)	For filing or later presentation of each claim (whether dependent or independent) in excess of 20	$20	$12	$10	$6
104 / 204	1.16(d)	If the application contains, or is amended to contain, a multiple dependent claim(s), per application	$200	$120	$100	$60
106 / 206	1.16(f)	For filing each design application	$250	$150	$125	$75
107 / 207	1.16(g)	Basic fee for filing an application for a plant patent	$420	$250	$210	$125
108 / 208	1.16(h)	Basic fee for filing an application for a reissue patent	$630	$370	$315	$185
109 / 209	1.16(i)	In a reissue application, for filing or later presentation of each independent claim in excess of independent claims in the original patent	$60	$36	$30	$18
110 / 210	1.16(j)	In a reissue application, for filing or later presentation of each claim (whether dependent or independent) in excess of 20 and also in excess of the number of claims in the original patent	$20	$12	$10	$6
115 / 215	1.17(a)	Extension fee for response within first month	$100	$62	$50	$31
116 / 216	1.17(b)	Extension fee for response within second month	$300	$180	$150	$90
117 / 217	1.17(c)	Extension fee for response within third month	$730	$430	$365	$215
118 / 218	1.17(d)	Extension fee for response within fourth month	$1,150	$680	$575	$340
119 / 219	1.17(e)	For filing a notice of appeal from the examiner to the Board of Patent Appeals and Interfer.	$240	$140	$120	$70
120 / 220	1.17(f)	In addition to the fee for filing a notice of appeal, for filing a brief in support of an appeal	$240	$140	$120	$70
121 / 221	1.17(g)	For filing a request for an oral hearing in an appeal under 35USC 134 before the BOPAI	$200	$120	$100	$60
140 / 240	1.17(l)	For filing a petition to the Commissioner (1) for revival of an abandoned application under 35 USC 133, or 371 or (2) for the delayed payment of the issue fee under 35 USC 151	$100	$62	$50	$31
141 / 241	1.17(m)	For filing a petition to the Commissioner (1) for revival of an unintentionally abandoned application or (2) for the unintentionally delayed payment of the fee for issuing a patent	$1,050	$620	$525	$310
142 / 242	1.18(a)	Issue fee for each original or reissue patent, except a design or plant patent	$1,050	$620	$525	$310
143 / 243	1.18(b)	Issue fee for a design patent	$370	$220	$185	$110
144 / 244	1.18(c)	Issue fee for a plant patent	$520	$310	$260	$155
148 / 248	1.20(d)	For filing each statutory disclaimer (§1.321)	$100	$62	$50	$31
173 / 273	1.20(h)	Maintaining original or reissue patent, except design or plant patent, in force beyond 4 years based on application filed on or after 8/27/82, due by 3.5 years after grant (P.L.97-247)	$830	$490	$415	$245
174 / 274	1.20(i)	Maintaining original or reissue patent, except design or plant patent, in force beyond 8 years based on application filed on or after 8/27/82, due by 7.5 years after grant (P.L.97-247)	$1,670	$990	$835	$495
175 / 275	1.20(j)	Maintaining original or reissue patent, except design or plant patent, in force beyond 12 years based on application filed on or after 8/27/82, due by 11.5 years after grant (P.L.97-247)	$2,500	$1,480	$1,250	$740

11/5/90

Appendix 5
Patent Fees as of June 30, 1987

Form PTO-442
(Rev. 3/88)

U.S. Department of Commerce
Patent and Trademark Office Fees
Effective June 30, 1987

Fee Code	Patent Processing Fees and Non-small Entity Fees	Fee
101	Basic fee for filing each application for an original patent, except design or plant cases (§1.16)	$340.00
102	In addition to the basic filing fee, for filing or later presentation of each independent claim in excess of 3	34.00
103	For filing or later presentation of each claim (whether independent or dependent) in excess of 20	12.00
104	If the application contains, or is amended to contain, a multiple dependent claim(s), per application	110.00
105	Surcharge for filing the basic filing fee or oath or declaration on a date later than the filing date of the application	110.00
106	For filing each design application	140.00
107	Basic fee for filing each plant application	220.00
108	Basic fee for filing each reissue application	340.00
109	In a reissue application, for filing or later presentation of each independent claim in excess of independent claims in the original patent	34.00
110	In a reissue application, for filing or later presentation of each claim (whether independent or dependent) in excess of 20 and also in excess of the number of claims in original patent	12.00
111	Filing an application for extension of the term of patent (35 U.S.C. 156)	750.00
112	For publication of a statutory invention registration prior to the mailing of the first examiner's action pursuant to §1.104, $400.00 reduced by the amount of the application basic filing fee previously paid.	--
	For publication of a statutory invention registration after the mailing of the first examiner's action pursuant to §1.104, $800.00 reduced by the amount of the application basic filing fee previously paid.	--

Fee Code	Patent Processing Fees and Non-small Entity Fees	Fee
115	Extension fee for response within first month pursuant to §1.136(a)	$ 56.00
116	Extension fee for response within second month pursuant to §1.136(a)	170.00
117	Extension fee for response within third month pursuant to §1.136(a)	390.00
118	Extension fee for response within fourth month pursuant to §1.136(a)	610.00
119	For filing a notice of appeal from the examiner to the Board of Patent Appeals and Interferences	130.00
120	In addition to the fee for filing a notice of appeal, for filing a brief in support of an appeal	130.00
121	For filing a request for an oral hearing in an appeal under 35 U.S.C. 134 before the Board of Patent Appeals and Interferences	110.00
122	For filing a petition to Commissioner for filing by other than all the inventors or a person not the inventor (§1.47)	140.00
123	For filing a petition to the Commissioner for correction of inventorship (§1.48)	140.00
124	For filing a petition to the Commissioner for decision on questions not specifically provided for (§1.182)	140.00
127	For filing a petition to the Commissioner for access to an assignment record (§1.12)	72.00
128	For filing a petition to the Commissioner for access to an application (§1.14)	72.00
129	For filing a petition to the Commissioner for entry of late priority papers (§1.55)	72.00
130	For filing a petition to the Commissioner to make application special (§1.102)	72.00
131	For filing a petition to the Commissioner to suspend action in application (§1.103)	72.00
132	For filing a petition to the Commissioner for divisional reissues to issue separately (§1.177)	72.00
133	For filing a petition to the Commissioner for access to interference settlement agreement (§1.666(b))	72.00
134	For filing a petition to the Commissioner for amendment after payment of issue fee (§1.312)	72.00
135	For filing a petition to the Commissioner to withdraw an application from issue (§1.313)	72.00
136	For filing a petition to the Commissioner to defer issuance of a patent (§1.314)	72.00
137	For filing a petition to the Commissioner for patent to issue to assignee, assignment recorded late (§1.334)	72.00
138	For filing a petition to institute a public use proceeding under §1.292	860.00
139	For processing an application filed with a specification in a non-English language (§1.52(d))	26.00

Fee Code	Patent Processing Fees and Non-small Entity Fees	Fee
140	For filing a petition (1) for revival of an abandoned application under 35 U.S.C. 133, or (2) for the delayed payment of the issue fee under 35 U.S.C. 151	$ 56.00
141	For filing a petition (1) for revival of an unintentionally abandoned application or (2) for the unintentionally delayed payment of the fee for issuing a patent	560.00
142	Issue fee for issuing each original or reissue patent, except a design or plant patent	560.00
143	Issue fee for issuing a design patent	200.00
144	Issue fee for issuing a plant patent	280.00
145	For providing a certificate of correction of applicant's mistake (§1.323)	29.00
146	Petition for correction of inventorship in patent (§1.324)	140.00
147	For filing a request for reexamination (§1.510(a))	1770.00
148	For filing each statutory disclaimer (§1.321)	56.00
150	PCT transmittal fee (see 35 U.S.C. 361(d) and PCT Rule 14)	170.00
151	PCT search fee where no corresponding prior U.S. National application with fee has been filed	520.00
152	Supplemental search fee when required by the U.S. PTO (see PCT Art. 17(3)(a) and PCT Rule 40.2) per additional invention	140.00
153	PCT search fee where corresponding prior U.S. national application with fee has been filed	000.00
154	Surcharge for filing the national fee or oath or declaration later than 20 months from the priority date	110.00
156	For filing an English translation of an international application later than 20 months after the priority date (§1.61(b))	26.00
159	PCT overpayments	--
160	Filing petition to Commissioner for expedited handling of foreign filing license (§5.12, §5.13, §5.14)	140.00
161	Filing petition to Commissioner for changing the scope of a license (§5.15)	140.00
162	Filing petition to Commissioner for retroactive license (§5.25)	140.00
163	Filing petition to Commissioner for review of decision refusing to accept and record payment of maintenance fee prior to expiration of patent (§1.377)	140.00
164	Filing petition to Commissioner for reconsideration of decision on petition refusing to accept delayed payment of maintenance fee in expired patent (§1.378(e))	140.00

Fee Code	Patent Processing Fees and Non-small Entity Fees	Fee
165	Filing petition to Commissioner-for petition in an interference (§1.644(e))	$140.00
166	Filing petition to Commissioner-for request for reconsideration of a decision on petition in an interference (§1.644(f))	140.00
167	Filing petition to Commissioner-for late filing of interference settlement agreement (§1.666(c))	140.00
168	Filing petition to Commissioner-for review of refusal to publish a statutory invention registration (§1.295)	140.00
170	*Maintaining original or reissue patent, except design or plant patent, in force beyond 4 years, based on applic. filed on or after 12/12/80 and before 8/27/82, due by 3½ yrs. after grant	225.00
171	*Maintaining original or reissue patent, except design or plant patent, in force beyond 8 years, based on application filed on or after 12/12/80 and before 8/27/82, due by 7½ yrs. after grant	445.00
172	*Maintaining original or reissue patent, except design or plant patent, in force beyond 12 years, based on application filed on or after 12/12/80 and before 8/27/82, due by 11½ yrs. after grant	670.00
173	Maintaining original or reissue patent, except design or plant patent, in force beyond 4 years, based on application filed on or after 8/27/82, due by 3½ yrs. after grant	450.00
174	Maintaining original or reissue patent, except design or plant patent, in force beyond 8 years, based on application filed on or after 8/27/82, due by 7½ yrs. after grant	890.00
175	Maintaining original or reissue patent, except design or plant patent, in force beyond 12 years, based on application filed on or after 8/27/82, due by 11½ yrs. after grant	1340.00
176	Surcharge for paying maintenance fee during 6 mos. grace period based on application filed on or after 12/12/80 and before 8/27/82	110.00
177	Surcharge for paying maintenance fee during 6 mos. grace period based on application filed on or after 8/27/82	110.00
178	Surcharge for accepting a maintenance fee after expiration of a patent for non-timely payment of a maintenance fee where the delay in payment is shown to satisfaction of the Commissioner to have been unavoidable	500.00
181	Maintenance fees received without sufficient information	--
190	Preliminary examination fee where an international search fee was not paid to the USPTO as International Searching Authority (ISA) in the international application (IA)	570.00
191	Preliminary examination fee where the International Searching Authority in the international application was the USPTO	370.00
192	Fee per additional invention when required where an international search fee was not paid to the USPTO as ISA in the IA	190.00
193	Fee per additional invention where the ISA in the IA was the USPTO	125.00

*Fee codes 170, 171, and 172 are not subject to small entity distinctions since public law 96-517 did not provide for small entities.

Fee Code	Patent Processing Fees—Small Entity	Fee
201	Basic application for original patent, except design or plant	$170.00
202	Each independent claim in excess of 3	17.00
203	Each claim in excess of 20 (whether independent or dependent)	6.00
204	Filing multiple dependent claims-per application	55.00
205	Surcharge on basic filing fee paid subsequent to application date	55.00
206	Filing each design application	70.00
207	Basic fee for filing plant application	110.00
208	Basic application for reissue patent	170.00
209	In reissue application-each independent claim in excess of no. in original patent	17.00
210	In reissue application-each claim (whether independent or dependent) in excess of 20, and also in excess of claims in original patent	6.00
215	Extension for response within first month	28.00
216	Extension for response within second month	85.00
217	Extension for response within third month	195.00
218	Extension for response within fourth month	305.00
219	Notice of appeal to Board of Appeals	65.00
220	Filing a brief in support of an appeal	65.00
221	Filing request for an oral hearing before Board of Appeals	55.00
240	Filing petition to revive abandoned application under 35 U.S.C. 133, or delayed payment of issue fee under 35 U.S.C. 151	28.00
241	Filing petition to revive an unintentionally abandoned application or unintentionally delayed payment of issue fee	280.00
242	Issue of original or reissue patent, except design or plant patent	280.00
243	Issue of design patent	100.00
244	Issue of plant patent	140.00

Appendix 6
Patent Applications Preserved in Secrecy

§1.14 Patent applications preserved in secrecy.

(a) Except as provided in §1.11(b) pending patent applications are preserved in secrecy. No information will be given by the Office respecting the filing by any particular person of an application for a patent, the pendency of any particular case before it, or the subject matter of any particular application, nor will access be given to or copies furnished of any pending application or papers relating thereto, without written authority in that particular application from the applicant or his assignee or attorney or agent of record, unless the application has been identified by serial number in a published patent document or the United States of America has been indicated as a Designated State in a published international application, in which case status information such as whether it is pending, abandoned or patented may be supplied, or unless it shall be necessary to the proper conduct of business before the Office or as provided by this part. Where an application has been patented, the patent number and issue date may also be supplied.

(b) Except as provided in §1.11(b) abandoned applications are likewise not open to public inspection, except that if an application referred to in a U.S. patent, or in an application in which the applicants has filed an authorization to open the complete application to the public, is abandoned and is available, it may be inspected or copies obtained by any person on written request, without notice to the applicant.

(c) Applications for patents which disclose, or which appear to disclose, or which purport to disclose, inventions or discoveries relating to atomic energy are reported to the Department of Energy, which Department will be given access to such applications, but such reporting does not constitute a determination that the subject matter of each application so reported is in fact useful or an invention or discovery or that such application in fact discloses subject matter in categories specified by sections 151(c) and 151(d) of the Atomic Energy Act of 1954, 68 Stat. 919; 42 U.S.C. 2181 (c) and (d).

(d) Any decision of the Board of Patent Appeals and Interferences, or any decision of the Commissioner on petition, not otherwise open to public inspection shall be published or made available for public inspection if: (1) The Commissioner believes the decision involves an interpretation of patent laws or regulations that would be of important precedent value; and (2) the applicant, or any party involved in the interference, does not within two months after being notified of the intention to make the decision public, object in writing on the ground that the decision discloses a trade secret or other confidential information. If a decision discloses such information, the applicant or party shall identify the deletions in the text of the decision considered necessary to protect the information. If it is considered the entire decision must be withheld from the public to protect such information, the applicant or party must explain why. Applicants or parties will be given time, not less than twenty days, to request reconsideration and seek court review before any portions of decisions are made public over their objection. See §2.27 for trademark applications.

(e) Any request by a member of the public seeking access to, or copies of, any pending or abandoned application preserved in secrecy pursuant to paragraphs (a) and (b) of this section, or any papers relating thereto, must

(1) Be in the form of a petition and be accompanied by the petition fee set forth in §1.17(i)(1), or

(2) Include written authority granting access to the member of the public in that particular application from the applicant or the applicant's assignee or attorney or agent of record.

NOTE: See §1.612(a) for access by an interference party to a pending or abandoned application.

(Pub. L. 94-131, 89 Stat. 685; 35 U.S.C. 6, Pub. L. 97-247; 15 U.S.C. 1113, 1123)

[42 FR 5593, Jan. 28, 1977, and 43 FR 20462, May 11, 1978, as amended at 49 FR 48451, Dec. 12, 1984; 50 FR 9378, Mar. 7, 1985; 53 FR 23733, June 23, 1988; 54 FR 6900, Feb. 15, 1989]

Appendix 7
Sample Page from Code of Federal Regulations—General Information and Correspondence

Subpart A—General Provisions

GENERAL INFORMATION AND CORRESPONDENCE

§1.1 All communications to be addressed to Commissioner of Patents and Trademarks.

(a) All letters and other communications intended for the Patent and Trademark Office must be addressed to "Commissioner of Patents and Trademarks," Washington, D.C. 20231. When appropriate, a letter should also be marked for the attention of a particular officer or individual.

(b) Letters and other communications relating to international applications during the international stage and prior to the assignment of a national serial number should be additionally marked "Box PCT."

(c) Requests for reexamination should be additionally marked "Box Reexam."

(d) Payments of maintenance fees in patents and other communications relating thereto should be additionally marked "Box M. Fee."

(e) Communications relating to interferences and applications or patents involved in an interference should be additionally marked "BOX INTERFERENCE."

(f) All applications for extension of patent term and any communications relating thereto intended for the Patent and Trademark Office should be additionally marked "Box Patent Ext." When appropriate, the communication should also be marked to the attention of a particular individual, as where a decision has been rendered.

(g) All communications relating to pending litigation which are required by the Federal Rules of Civil or Appellate Procedure or by a rule or order of a court to be served on the Solicitor shall be hand-delivered to the Office of the Solicitor or shall be mailed to: Office of the Solicitor, P.O. Box 15667, Arlington, Virginia 22215 or such other address as may be designated in writing in the litigation. All other communications to the Office of the Solicitor should be addressed to: Box 8, Commissioner of Patents and Trademarks, Washington, D.C. 20231. Any communication which does not involve pending litigation which is received at P.O. Box 15667 will not be filed in the Office but will be returned. See §§1.302(c) and 2.145(b)(3) for filing a notice of appeal to the U.S. Court of Appeals for the Federal Circuit.

NOTE: Sections 1.1 to 1.26 are applicable to trademark cases as well as to national and international patent cases except for provisions specifically directed to patent cases. See §1.9 for definitions of "national application" and "international application."

(Pub. L. 94-131, 89 Stat. 685)

[46 FR 29181, May 29, 1981, as amended at 49 FR 34724, Aug. 31, 1984; 49 FR 48451, Dec. 12, 1984; 52 FR 9394, Mar. 24, 1987; 53 FR 16413, May 9, 1988]

Appendix 8
Patent Application Declaration Form

OMB No. 0651-0011 (12/31/86)
DECLARATION FOR PATENT APPLICATION

Docket No. ________

As a below named inventor, I hereby declare that:

My residence, post office address and citizenship are as stated below next to my name.

I believe I am the original, first and sole inventor (if only one name is listed below) or an original, first and joint inventor (if plural names are listed below) of the subject matter which is claimed and for which a patent is sought on the invention entitled ________________, the specification of which

(check one) ☐ is attached hereto.
☐ was filed on ________________ as
Application Serial No. ________________
and was amended on ________________ (if applicable).

I hereby state that I have reviewed and understand the contents of the above identified specification, including the claims, as amended by any amendment referred to above.

I acknowledge the duty to disclose information which is material to the examination of this application in accordance with Title 37, Code of Federal Regulations, §1.56(a).

I hereby claim foreign priority benefits under Title 35, United States Code, §119 of any foreign application(s) for patent or inventor's certificate listed below and have also identified below any foreign application for patent or inventor's certificate having a filing date before that of the application on which priority is claimed:

Prior Foreign Application(s)

			Priority Claimed	
________ (Number)	________ (Country)	________ (Day/Month/Year Filed)	Yes	No
________ (Number)	________ (Country)	________ (Day/Month/Year Filed)	Yes	No

I hereby claim the benefit under Title 35, United States Code, §120 of any United States application(s) listed below and, insofar as the subject matter of each of the claims of this application is not disclosed in the prior United States application in the manner provided by the first paragraph of Title 35, United States Code, §112, I acknowledge the duty to disclose material information as defined in Title 37, Code of Federal Regulations, §1.56(a) which occurred between the filing date of the prior application and the national or PCT international filing date of this application:

________ (Application Serial No.)	________ (Filing Date)	________ (Status—patented, pending, abandoned)
________ (Application Serial No.)	________ (Filing Date)	________ (Status—patented, pending, abandoned)

I hereby appoint the following attorney(s) and/or agent(s) to prosecute this application and to transact all business in the Patent and Trademark Office connected therewith:
________________.

Address all telephone calls to ________________ at telephone no. ________________.
Address all correspondence to ________________
____

I hereby declare that all statements made herein of my own knowledge are true and that all statements made on information and belief are believed to be true; and further that these statements were made with the knowledge that willful false statements and the like so made are punishable by fine or imprisonment, or both, under Section 1001 of Title 18 of the United States Code and that such willful false statements may jeopardize the validity of the application or any patent issued thereon.

Full name of sole or first inventor ________________
Inventor's signature ________________ Date ________________
Residence ________________ Citizenship ________________
Post Office Address ________________

Full name of second joint inventor, if any ________________
Second Inventor's signature ________________ Date ________________
Residence ________________ Citizenship ________________
Post Office Address ________________

(Supply similar information and signature for third and subsequent joint inventors.)

Form PTO-FB-A110 (8-83)

Appendix 9
Patent Application Form—Declaration of Independent Inventor

OMB No. 0651-0011 (12/31/86)

Applicant or Patentee: ______________________ Attorney's
Serial or Patent No.: ______________________ Docket No.: ______
Filed or Issued: ______________________
For: ______________________

VERIFIED STATEMENT (DECLARATION) CLAIMING SMALL ENTITY
STATUS (37 CFR 1.9 (f) and 1.27 (b))—INDEPENDENT INVENTOR

As a below named inventor, I hereby declare that I qualify as an independent as defined in 37 CFR 1.9 (c) for purposes of paying reduced fees under Section 41 (a) and (b) of Title 35, United States Code, to the Patent and Trademark Office with regard to the invention entitled ______________________ described in

[ ] the specification filed herewith
[ ] application serial no. ______________________, filed ______________________.
[ ] patent no. ______________________, issued ______________________.

I have not assigned, granted, conveyed or licensed and am under no obligation under contract or law to assign, grant, convey or license, any rights in the invention to any person who could not be classified as an independent inventor under 37 CFR 1.9 (c) if that person had made the invention, or to any concern which would not qualify as a small business concern under 37 CFR 1.9 (d) or a nonprofit organization under 37 CFR 1.9 (e).

Each person, concern or organization to which I have assigned, granted, conveyed, or licensed or am under an obligation under contract or law to assign, grant, convey, or license any rights in the invention is listed below:

[ ] no such person, concern, or organization
[ ] persons, concerns or organizations listed below*

*NOTE: Separate verified statements are required from each named person, concern or organization having rights to the invention averring to their status as small entities. (37 CFR 1.27)

FULL NAME ______________________
ADDRESS ______________________
[ ] INDIVIDUAL [ ] SMALL BUSINESS CONCERN [ ] NONPROFIT ORGANIZATION

FULL NAME ______________________
ADDRESS ______________________
[ ] INDIVIDUAL [ ] SMALL BUSINESS CONCERN [ ] NONPROFIT ORGANIZATION

FULL NAME ______________________
ADDRESS ______________________
[ ] INDIVIDUAL [ ] SMALL BUSINESS CONCERN [ ] NONPROFIT ORGANIZATION

I acknowledge the duty to file, in this application or patent, notification of any change in status resulting in loss of entitlement to small entity status prior to paying, or at the time of paying, the earliest of the issue fee or any maintenance fee due after the date on which status as a small entity is no longer appropriate. (37 CFR 1.28 (b))

I hereby declare that all statement made herein of my own knowledge are true and that all statements made on information and belief are believed to be true; and further that these statements were made with the knowledge that willful false statements and the like so made are punishable by fine or imprisonment, or both, under section 1001 of Title 18 of the United States Code, and that such willful false statements may jeopardize the validity of the application, any patent issuing thereon, or any patent to which this verified statement is directed.

NAME OF INVENTOR	NAME OF INVENTOR	NAME OF INVENTOR
Signature of Inventor	Signature of Inventor	Signature of Inventor
Date	Date	Date

Form PTO-FB-A410 (8-83)

Appendix 10
Statutory Invention Registration

United States Statutory Invention Registration [19]

[11] Reg. Number: H194

Oakley

[43] Published: Jan. 6, 1987

[54] DECONTAMINATION APPARATUS AND METHOD

[75] Inventor: David J. Oakley, Richland, Wash.

[73] Assignee: The United States of America as represented by the United States Department of Energy, Washington, D.C.

[21] Appl. No.: 562,148

[22] Filed: Dec. 16, 1983

[51] Int. Cl.[4] G21C 19/42

[52] U.S. Cl. 376/310; 376/316; 134/199

[58] Field of Search .... 376/310, 316; 134/199, 134/122 R; 68/5 E; 34/242; 432/242

[56] References Cited

U.S. PATENT DOCUMENTS

860,199	7/1907	Emerson	
1,910,497	5/1933	Peik	134/199 X
4,064,582	12/1977	Sando et al.	34/242 X
4,296,556	10/1981	Bray	
4,401,619	8/1983	McEdwards	376/310 X
4,461,650	7/1984	Ozawa	376/310 X

FOREIGN PATENT DOCUMENTS

817750 3/1981 U.S.S.R.

OTHER PUBLICATIONS

Nielsen et al., "Wash and Dry Apparatus" IBM Technical Disclosure Bulletin, vol. 18, No. 5, (Oct. 1975), pp. 1349–1350.

Primary Examiner—Deborah L. Kyle
Assistant Examiner—John S. Maples
Attorney, Agent, or Firm—Robert Southworth, III; Judson R. Hightower

[57] ABSTRACT

A blast head including a plurality of spray nozzles mounted in a chamber for receiving a workpiece. The several spray nozzles concurrently direct a plurality of streams of a pressurized gas and abrasive grit mixture toward a peripheral portion of the workpiece to remove particulates or debris therefrom. An exhaust outlet is formed in the chamber for discharging the particulates and spent grit.

11 Claims, 5 Drawing Figures

A statutory invention registration is not a patent. It has the defensive attributes of a patent but does not have the enforceable attributes of a patent. No article or advertisement or the like may use the term patent, or any term suggestive of a patent, when referring to a statutory invention registration. For more specific information on the rights associated with a statutory invention registration see 35 U.S.C. 157.

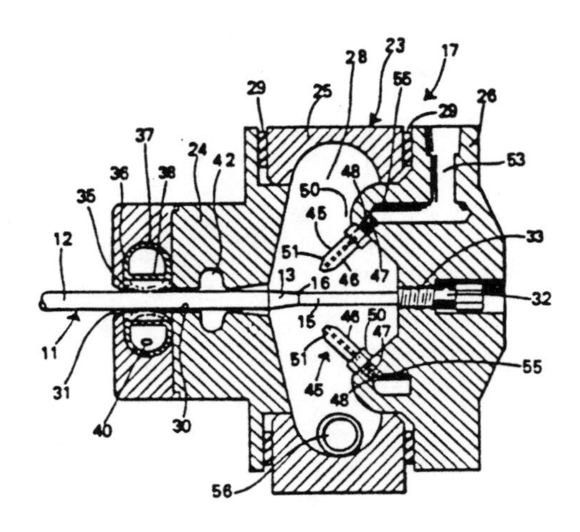

Appendix 11
Sample Pages from *Manual of Patent Examining Procedure*

Manual of PATENT EXAMINING PROCEDURE

Original Fifth Edition, August 1983
Latest Revision May 1988

U.S. DEPARTMENT OF COMMERCE
Patent and Trademark Office

Rev. 8, May 1988

608.01 Specification [R-8]

35 U.S.C. 22. Printing of papers filed.

The Commissioner may require papers filed in the Patent and Trademark Office to be printed or typewritten.

37 CFR 1.71 Detailed description and specification of the invention.

(a) The specification must include a written description of the invention or discovery and of the manner and process of making and using the same, and is required to be in such full, clear, concise, and exact terms as to enable any person skilled in the art or science to which the invention or discovery appertains, or with which it is most nearly connected, to make and use the same.

(b) The specification must set forth the precise invention for which a patent is solicited, in such manner as to distinguish it from other inventions and from what is old. It must describe completely a specific embodiment of the process, machine, manufacture, composition of matter or improvement invented, and must explain the mode of operation or principle whenever applicable. The best mode contemplated by the inventor of carrying out his invention must be set forth.

(c) In the case of an improvement, the specification must particularly point out the part or parts of the process, machine, manufacture, or composition of matter to which the improvement relates, and the description should be confined to the specific improvement and to such parts as necessarily cooperate with it or as may be necessary to a complete understanding or description of it.

Certain cross notes to other related applications may be made. References to foreign applications or to applications identified only by the attorney's docket number should be required to be cancelled. See 37 CFR 1.78 and MPEP §202.01.

37 CFR 1.52. Language, paper, writing, margins.

(a) The application, any amendments or corrections thereto, and the oath or declaration must be in the English language except as provided for in §1.69 and paragraph (d) of this section, or be accompanied by a verified translation of the application and a translation of any corrections or amendments into the English language. All papers which are to become a part of the permanent records of the Patent and Trademark Office must be legibly written, typed, or printed in permanent ink or its equivalent in quality. All of the application papers must be presented in a form having sufficient clarity and contrast between the paper and the writing, typing, or printing thereon to permit the direct production of readily legible copies in any number by use of photographic, electrostatic, photo-offset, and microfilming processes. If the papers are not of the required quality, substitute typewritten or printed papers of suitable quality may be required.

(b) The application papers (specification, including claims, abstract, oath or declaration, and papers as provided for in §§1.42, 1.43, 1.47, etc.) and also papers subsequently filed, must be plainly written on but one side of the paper. The size of all sheets of paper should be 8 to 8½ by 10½ to 13 inches (20.3 to 21.6 cm. by 26.6 to 33.0 cm.) A margin of at least approximately 1 inch (2.5 cm.) must be reserved on the left-hand of each page. The top of each page of the application, including claims must have a margin of at least approximately ¾ inch (2 cm.). The lines must not be crowded too closely together; typewritten lines should be 1½ or double spaced. The pages of the application including claims and abstract should be numbered consecutively, starting with 1, the numbers being centrally located above or preferably, below, the text.

(c) Any interlineation, erasure, or cancellation or other alteration of the application papers filed must be made before the signing of any accompanying oath or declaration pursuant to §1.63 referring to those application papers and should be dated and initialed or signed by the applicant on the same sheet of paper. No such alterations in the application papers are permissible after the signing of an oath or declaration referring to those application papers (§1.56(c)). After the signing of the oath or declaration referring to the application papers, amendments may only be made in the manner provided by §§1.121 and 1.123 through 1.125.

(d) An application ** may be filed in a language other than English**. A verified English translation of the non-English language application and the fee set forth in §1.17(k) are * required to be filed with the application or within such time as may be set by the Office.

37 CFR 1.58 Chemical and mathematical formulas and tables.

(a) The specification, including the claims, may contain chemical and mathematical formulas, but shall not contain drawings or flow diagrams. The description portion of the specification may contain tables; claims may contain tables only if necessary to conform to 35 U.S.C. 112 or if otherwise found to be desirable.

(b) All tables and chemical and mathematical formulas in the specification, including claims, and amendments thereto, must be on paper which is flexible, strong, white, smooth, nonshiny, and durable, in order to permit use as camera copy when printing any patent which may issue. A good grade of bond paper is acceptable; watermarks should not be prominent. India ink or its equivalent, or solid black typewriter should be used to secure perfectly black solid lines.

(c) To facilitate camera copying when printing, the width of formulas and tables as presented should be limited normally to 5 inches (12.7 cm.) so that it may appear as a single column in the printed patent. If it is not possible to limit the width of a formula or table to 5 inches (12.7 cm.), it is permissible to present the formula or table with a maximum width of 10¾ inches (27.3 cm.) and to place it sideways on the sheet. Typewritten characters used in such formulas and tables must be from a block (nonscript) type font or lettering style having capital letters which are at least 0.08 inch (2.1 mm.) high (elite type). Hand lettering must be neat, clean, and have a minimum character height of 0.08 inch (2.1 mm.). A space at least ¼ inch (6.4 mm.) high should be provided between complex formulas and tables and the text. Tables should have the lines and columns of data closely spaced to conserve space, consistent with high degree of legibility.

Appendix 12
Reissue Patent

REISSUES

NOVEMBER 17, 1987

Matter enclosed in heavy brackets [ ] appears in the original patent but forms no part of this reissue specification; matter printed in italics indicates additions made by reissue.

Re. 32,545
MAGNETIC DOMAIN LOGIC DEVICE
Jan W. F. Dorleijn; Willem F. Druyvesteyn, and Frederik A. De Jonge, all of Eindhoven, Netherlands, assignors to U.S. Philips Corporation, New York, N.Y.
Original No. 3,944,842, dated Mar. 16, 1976, Ser. No. 478,575, Jun. 12, 1974. Continuation of Ser. No. 277,150, Aug. 2, 1972, abandoned. Application for reissue Oct. 20, 1977, Ser. No. 844,109
Claims priority, application Netherlands, Aug. 3, 1971, 7110674
Int. Cl.⁴ G11C *19/08*
U.S. Cl. 365—32 13 Claims

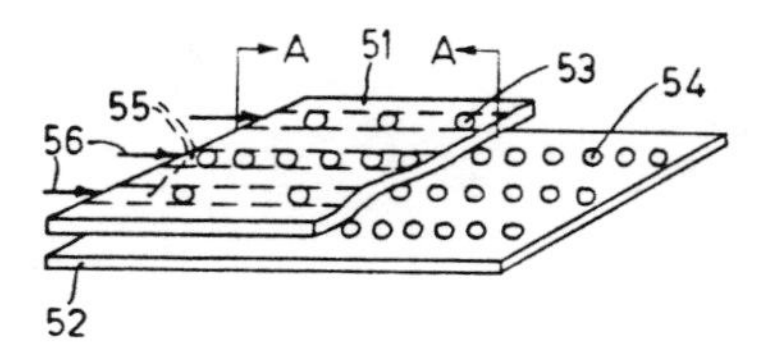

4. A magnetic device comprising a first plate of a magnetic material in which at least one domain is present, a domain guiding structure having a plurality of guiding paths for the displacement of domains in the first plate, and a second plate of magnetic material completely filled with domains, said second domain covering, at least in projection, at least a part of the first plate, an interaction force occurring between at least one domain in the first and one domain in the second plate, the domains in the second plate creating stable positions for the domains in the first plate.

1001

Appendix 13
Patent Depository Libraries

PATENT EXAMINING CORPS

RENE D. TEGTMEYER, Assistant Commissioner
JAMES E. DENNY, Deputy Assistant Commissioner

CONDITION OF PATENT APPLICATIONS AS OF October 10, 1987

PATENT EXAMINING GROUPS	Actual Filing Date of Oldest New Case Awaiting Action
CHEMICAL EXAMINING GROUPS	
GENERAL METALLURGICAL, INORGANIC, PETROLEUM AND ELECTRICAL CHEMISTRY, AND ENGINEERING, GROUP 110—D. E. TALBERT, Director	8-05-86
ORGANIC CHEMISTRY AND BIOTECHNOLOGY, GROUP 120—C. E. VAN HORN, Director	2-15-85
SPECIALIZED CHEMICAL INDUSTRIES AND CHEMICAL ENGINEERING, GROUP 130—R. F. WHITE, Director	1-27-87
HIGH POLYMER CHEMISTRY, PLASTICS, COATING, PHOTOGRAPHY, STOCK MATERIALS AND COMPOSITIONS, GROUP 150—J. O. THOMAS, Director	1-29-86
ELECTRICAL EXAMINING GROUPS	
INDUSTRIAL ELECTRONICS, PHYSICS AND RELATED ELEMENTS, GROUP 210—G. GOLDBERG, Director	2-18-86
SPECIAL LAWS ADMINISTRATION, GROUP 220—K. L. CAGE, Director	10-14-85
INFORMATION PROCESSING, STORAGE, AND RETRIEVAL, GROUP 230—E. LEVY, Director	1-07-85
PACKAGES, CLEANING, TEXTILES, AND GEOMETRICAL INSTRUMENTS, GROUP 240—TRYGVE M. BLIX, Director	5-21-85
ELECTRONIC AND OPTICAL SYSTEMS AND DEVICES, GROUP 250—EDWARD E. KUBASIEWICZ, Director	9-06-85
COMMUNICATIONS, MEASURING, TESTING AND LAMP/DISCHARGE GROUP, GROUP 260—S. G. KUNIN, Director	2-20-86
DESIGN, GROUP 290—K. L. CAGE, Director	1-18-85
MECHANICAL EXAMINING GROUPS	
HANDLING AND TRANSPORTING MEDIA, GROUP 310—B. R. GRAY, Director	9-12-86
MATERIAL SHAPING, ARTICLE MANUFACTURING AND TOOLS, GROUP 320—S. N. ZAHARNA, Director	10-04-85
MECHANICAL TECHNOLOGIES AND HUSBANDRY PERSONAL TREATMENT INFORMATION, GROUP 330—R. E. AEGERTER, Director	10-15-85
SOLAR, HEAT, POWER, AND FLUID ENGINEERING DEVICES, GROUP 340—D. J. STOCKING, Director	12-29-86
GENERAL CONSTRUCTIONS, PETROLEUM AND MINING ENGINEERING, GROUP 350—A. L. SMITH, Director	1-02-87

Expiration of patents: The patents within the range of numbers indicated below expire during October 1987, except those which may have had their terms curtailed by disclaimer under the provisions of 35 U.S.C. 253. Other patents, issued after the dates of the range of numbers indicated below, may have expired before the full term of 17 years for the same reasons, or have lapsed under the provisions of 35 U.S.C. 151.

Patents . . . . . Numbers 3,531,806 to 3,537,106, inclusive
Plant Patents . . . . . **Numbers 2,991 to 2,995 inclusive**

1084 OG 26

Reference Collections of U.S. Patents Available for Public Use in Patent Depository Libraries

The following libraries, designated as Patent Depository Libraries, receive current issues of U.S. Patents and maintain collections of earlier issued patents. The scope of these collections varies from library to library, ranging from patents of only recent years to all or most of the patents issued since 1790.

These patent collections are open to public use and each of the Patent Depository Libraries, in addition, offers the publications of the U.S. Patent Classification System (e.g. The Manual of Classification, Index to the U.S. Patent Classification, Classification Definitions, etc.) and provides technical staff assistance in their use to aid the public in gaining effective access to information contained in patents. With one exception, as noted in the table following, the collections are organized in patent number sequence.

Facilities for making paper copies from either microfilm in reader-printers or from the bound volumes in paper-to-paper copies are generally provided for a fee.

Owing to variations in the scope of patent collections among the Patent Depository Libraries and in their hours of service to the public, anyone contemplating use of the patents at a particular library is advised to contact that library, in advance, about its collection and hours, so as to avert possible inconvenience.

State	*Name of Library*	*Telephone Contact*
Alabama	Auburn University Libraries	(205) 826-4500 Ext. 21
	Birmingham Public Library	(205) 226-3680
Alaska	Anchorage Municipal Libraries	(907) 261-2907
Arizona	Tempe: Noble Library, Arizona State University	(602) 965-7140
Arkansas	Little Rock: Arkansas State Library	(501) 371-2090
California	Irvine: University of California, Irvine Library	(714) 856-7234
	Los Angeles Public Library	(213) 612-3273
	Sacramento: California State Library	(916) 322-4572
	San Diego Public Library	(619) 236-5813
	Sunnyvale: Patent Information Clearinghouse*	(408) 730-7290
Colorado	Denver Public Library	(303) 571-2347
Connecticut	New Haven: Science Park Library	(203) 786-5447
Delaware	Newark: University of Delaware Library	(302) 451-2965
Dist. of Columbia	Washington: Howard University Libraries	(202) 636-5060
Florida	Fort Lauderdale: Broward County Main Library	(305) 357-7444
	Miami-Dade Public Library	(305) 375-2665
Georgia	Atlanta: Price Gilbert Memorial Library, Georgia Institute of Technology	(404) 894-4508
Idaho	Moscow: University of Idaho Library	(208) 885-6235
Illinois	Chicago Public Library	(312) 269-2865
	Springfield: Illinois State Library	(217) 782-5430
Indiana	Indianapolis-Marion County Public Library	(317) 269-1741
Louisiana	Baton Rouge: Troy H. Middleton Library, Louisiana State University	(504) 388-2570
Maryland	College Park: Engineering and Physical Sciences Library, University of Maryland	(301) 454-3037
Massachusetts	Amherst: Physical Sciences Library, University of Massachusetts	(413) 545-1370
	Boston Public Library	(617) 536-5400 Ext. 265
Michigan	Ann Arbor: Engineering Transportation Library, University of Michigan	(313) 764-7494
	Detroit Public Library	(313) 833-1450
Minnesota	Minneapolis Public Library & Information Center	(612) 372-6570
Missouri	Kansas City: Linda Hall Library	(816) 363-4600
	St. Louis Public Library	(314) 241-2288 Ext. 390
Montana	Butte: Montana College of Mineral Science and Technology Library	(406) 496-4222
Nebraska	Lincoln: University of Nebraska-Lincoln, Engineering Library	(402) 472-3411
Nevada	Reno: University of Nevada Library	(702) 784-6579
New Hampshire	Durham: University of New Hampshire Library	(603) 862-1777
New Jersey	Newark Public Library	(201) 733-7815
New Mexico	Albuquerque: University of New Mexico Library	(505) 277-5441
New York	Albany: New York State Library	(518) 474-7040
	Buffalo and Erie County Public Library	(716) 846-7101
	New York Public Library (The Research Libraries)	(212) 714-8529
North Carolina	Raleigh: D. H. Hill Library, N.C. State University	(919) 737-3280
Ohio	Cincinnati & Hamilton County, Public Library of	(513) 369-6936
	Cleveland Public Library	(216) 623-2870
	Columbus: Ohio State University Libraries	(614) 292-6286
	Toledo/Lucas County Public Library	(419) 255-7055 Ext. 212
Oklahoma	Stillwater: Oklahoma State University Library	(405) 624-6546
Oregon	Salem: Oregon State Library	(503) 378-4239
Pennsylvania	Philadelphia: Free Library	(215) 686-5330
	Pittsburgh: Carnegie Library of Pittsburgh	(412) 622-3138
	University Park: Pattee Library, Pennsylvania State University	(814) 865-4861
Rhode Island	Providence Public Library	(401) 521-8726
South Carolina	Charleston: Medical University of South Carolina Library	(803) 792-2371
Tennessee	Memphis & Shelby County Public Library and Information Center	(901) 725-8876
	Nashville: Vanderbilt University Library	(615) 322-2775
Texas	Austin: McKinney Engineering Library, University of Texas.	(512) 471-1610
	College Station: Sterling C. Evans Library, Texas A & M University	(409) 845-2551
	Dallas Public Library	(214) 670-1468
	Houston: The Fondren Library, Rice University	(713) 527-8101 Ext. 2587
Utah	Salt Lake City: Marriott Library, University of Utah	(801) 581-8394
Virginia	Richmond: Virginia Commonwealth University Library	(804) 257-1104
Washington	Seattle: Engineering Library, University of Washington	(206) 543-0740
Wisconsin	Madison: Kurt F. Wendt Library, University of Wisconsin	(608) 262-6845
	Milwaukee Public Library	(414) 278-3247

All of the above-listed libraries offer CASSIS (Classification And Search Support Information System), which provides direct, on-line access to Patent and Trademark Office data.

*Collection organized by subject matter.

Appendix 14
Sample Page from *U.S. Serial Set, Report of the Commissioner of Patents*

JAMES MONROE.

To the Hon. the SPEAKER *of the House of Representatives of the United States.*

LIST OF PATENTEES.

Inventions.	When issued.	Names of patentees.	Residence.
A nurse lamp,	1812. Dec. 31	William Howe,	Boston.
A pendulum mill,	1813. Jan. 6	William Grandin,	Hector, Seneca co. N. Y.
A machine for making ship and other bread,	8	Jehosaphat Starr,	Middletown, Middlesex co. Conn.
In pitchforks,	12	Jared Byington,	Hinesborough, Chittenden co. Vermont.
In preparing colors and paints from ores, &c.	14	Henry Alexander,	Baltimore.
In the fireplace and chimney for saving fuel,	18	Samuel Morey,	Philadelphia.
In distilling,	18	Omri Carrier,	Enfield, Connecticut.
In distilling,	26	John Bates,	Hartford, Connecticut.
In making cloth and blankets from sheep's wool,	Feb. 3	Allen Barnes, Samuel Gray, and Jabez Clark.	Windham, Connecticut.
A machine for cutting fur from peltry,	4	Ephraim Cutter,	Walpole, New Hampshire.
A machine for heading wood screws,	4	Abel Stowell,	Worcester, Worcester co. Massachusetts.
Manufacturing acetate of copper or verdigris,	4	Stephen Dempsey,	New York.
In the stove and boiler,	6	Gabriel N. Phillips,	Goshen, Orange co. N. Y.
In raising water from wells, &c.	6	George Patterson,	Philadelphia.
A machine for carding cloth,	6	John Jessup,	Orange county, New York.
In the kitchen stove,	8	John Spencer,	Albany, New York.
Machine for making bread,	8	George Richards,	Manlius, Onondaga co. N. Y.
In the plough,	8	John Sietz,	Strasburg, Lancaster co. Pen.
Wheels of elastic oars for propelling boats, &c.	16	Louis Mure Latour.	
A printing press and mode of distributing ink on types,	17	William Elliot,	New York.
In the printing press,	26	Zachariah Mills,	Hartford, Connecticut.
In tassels and cords for boots,	March 1	Alfred Janes,	Hartford, Connecticut.
In stoves,	1	Alfred Janes,	Hartford, Connecticut.
Machine for planing boards and plank, and shaving shingles,	2	William Badger,	Madison county, Miss'pi T.
An ice-breaking machine,	2	John I. Williams,	Georgetown, Dist. Columbia.
In saw mills,	2	Hiram Whitcomb,	Cornwall, Litchfield co. Con.
A medicine made use of in fevers, &c.	2	Samuel Thompson,	Surry, Cheshire co. N. H.
For boiling liquors and for distillation,	3	John Morris,	New Haven, Connecticut.
The perpetual wool spinner,	3	Burgiss Allison,	Burlington, New Jersey.
For sifting meal,	8	Daniel Homan,	Brookhaven, Suffolk co. N.Y.
A refrigerator,	16	John W. Bronaugh and Jesse Talbot,	Georgetown, Dist. Columbia.
In manufacturing brass, copper, and composition nails,	17	George Whitfield Robinson,	Attleborough, Bristol county, Massachusetts.
Mode of applying heat to machines for preparing wool for worsted,	18	Joseph Bamford,	Philadelphia.
In the wheel-head,	18	Eli Church,	Homer, Cortland co. N. Y.
In fireplaces,	18	David F. Launy,	Philadelphia.
In the gas lamp,	18	David Melville,	Newport, Rhode Island.
A domestic and factory loom,	20	Walter Janes,	Ashford, Windham co. Con.
In repeating gunnery,	23	Joseph C. Chambers,	West Middletown, Washington county, Pennsylvania.
In pumps,	23	Jacob Perkins,	Newburyport, Massach'setts.
Machine for making screw shanks,	23	Jacob Perkins,	Newburyport, Massach'setts.
In the fire engine,	23	Jacob Perkins,	Newburyport, Massach'setts.
A vault lock for banks,	23	Jacob Perkins,	Newburyport, Massach'setts.
A portable or separating furnace for casting brass or other metal,	30	Joseph Share,	Baltimore.
In the boiling stoves,	30	Henry Abbott,	Philadelphia.
In stoves,	30	George Warrall,	Philadelphia.
For shaving and splitting leather,	April 5	Samuel Parker,	Billerica, Middlesex co. Mas.
In the mode of preparing magnesia,	8	William Dunn,	Boston.

Appendix 15
Sample Patent from People's Republic of China

〔19〕中华人民共和国专利局

〔51〕Int Cl.4
B01J 23/80
B01J 37/02
C07C 121/54

〔12〕发明专利申请公开说明书

〔11〕CN 85 1 01963 A *

CN 85 1 01963 A

〔43〕公开日 1986年9月3日

〔21〕申请号 85 1 01963
〔22〕申请日 85．4．1
〔71〕申请人 云南省化工研究所
地址 云南省昆明市东风东路
〔72〕发明人 林恒兴 彭明楷 马瑜华 王绍存

〔74〕专利代理机构 云南省专利事务所
代理人 彭远东

〔54〕发明名称 制备四氯间苯二甲腈的催化剂制法

〔57〕摘要

采用浸煮干燥法，可按比例、定量地将数种活性组分载附于同一载体上。如：将氯化铁、氯化铜、氯化锌等三种活性组分，按比例溶解于稀盐酸中，加入活性炭载体，浸煮至干，再干燥后即得复合催化剂。制备的复合催化剂，在间苯二甲腈制备四氯间苯二甲腈时，在300℃高温氯化反应中，具有反应缓和、氯化转化率高（98%以上），每公斤催化剂可生产产品250～300公斤等优点。

北京市期刊登记证第1405号

Appendix 16
Rules for Drawings from the Code of Federal Regulations

§1.84 Standards for drawings.

(a) *Paper and ink*. Drawings or high quality copies thereof which are submitted to the Office must be made upon paper which is flexible, strong, white, smooth, non-shiny and durable. India ink, or its equivalent in quality, is preferred for pen drawings to secure perfectly black solid lines. The use of white pigment to cover lines is not normally acceptable. See paragraph (p) of this section for use of color drawings in utility patent applications.

(b) *Size of sheet and margins*. The size of the sheets on which drawings are made may be exactly 8½ by 14 inches (21.6 by 35.6 cm.), exactly 8½ by 13 inches (21.6 by 33.1 cm.), or exactly 21.0 by 29.7 cm. (DIN size A4). All drawing sheets in a particular application must be the same size. One of the shorter sides of the sheet is regarded as its top.

(1) On 8½ by 14 inch drawing sheets, the drawing must include a top margin of 2 inches (5.1 cm.) and bottom and side margins of ¼ inch (6.4 mm.) from the edges, thereby leaving a "sight" precisely 8 by 11¾ inches (20.3 by 29.8 cm.). Margin border lines are not permitted. All work must be included within the "sight." The sheets may be provided with two ¼ inch (6.4 mm.) diameter holes having their centerlines spaced 11/16 inch (17.5 mm.) below the top edge and 2¾ inches (7.0 cm.) apart, said holes being equally spaced from the respective side edges.

(2) On 8½ by 13 inch drawing sheets, the drawing must include a top margin of 1 inch (2.5 cm.) and bottom and side margins of ¼ inch (6.4 mm.) from the edges, thereby leaving a "sight" precisely 8 by 11¾ inches (20.3 by 29.8 cm.). Margin border lines are not permitted. All work must be included within the "sight." The sheets may be provided with two ¼ inch (6.4 mm.) diameter holes having their centerlines spaced 11 11/16 inch (17.5 mm.) below the top edge and 2¾ inches (7.0 cm.) apart, said holes being equally spaced from the respective side edges.

(3) On 21.0 by 29.7 cm. drawing sheets, the drawing must include a top margin of at least 2.5 cm., a left side margin of 2.5 cm., a right side margin of 1.5 cm., and a bottom margin of 1.0 cm. Margin border lines are not permitted. All work must be contained within a sight size not to exceed 17 by 26.2 cm.

(c) *Character of lines*. All drawings must be made with drafting instruments or by a process which will give them satisfactory reproduction characteristics. Every line and letter must be durable, black, sufficiently dense and dark, uniformly thick and well defined; the weight of all lines and letters must be heavy enough to permit adequate reproduction. This direction applies to all lines however fine, to shading, and to lines representing cut surfaces in sectional views. All lines must be clean, sharp, and solid. Fine or crowded lines should be avoided. Solid black should not be used for sectional or surface shading. Freehand work should be avoided wherever it is possible to do so.

(d) *Hatching and shading*. (1) Hatching should be made by oblique parallel lines spaced sufficiently apart to enable the lines to be distinquished without difficulty.

(2) Heavy lines on the shade side of objects should preferably be used except where they tend to thicken the work and obscure reference characters. The light should come from the upper left-hand corner at an angle of 45°. Surface delineations should preferably be shown by proper shading, which should be open.

(e) *Scale*. The scale to which a drawing is made ought to be large enough to show the mechanism without crowding when the drawing is reduced in size to two-thirds in reproduction, and views of portions of the mechanism on a larger scale should be used when necessary to show details clearly; two or more sheets should be used if one does not give sufficient room to accomplish this end, but the number of sheets should not be more than is necessary.

(f) *Reference characters*. The different views should be consecutively numbered figures. Reference numerals (and letters, but numerals are preferred) must be plain, legible and carefully formed, and not be encircled. They should, if possible, measure at least one-eighth of an inch (3.2 mm.) in height so that they may bear reduction to one twenty-fourth of an inch (1.1 mm.); and they may be slightly larger when there is sufficient room. They should not be so placed in the close and complex parts of the drawing as to interfere with a thorough comprehension of the same, and therefore should rarely cross or mingle with the lines. When necessarily grouped around a certain part, they should be placed at a little distance, at the closest point where there is available space, and connected by lines with the parts to which they refer. They should not be placed upon hatched or shaded surfaces but when necessary, a blank space may be left in the hatching or shading where the character occurs so that it shall appear perfectly distinct and separate from the work. The same part of an invention appearing in more than one view of the drawing must always be designated by the same character, and the same character must never be used to designate different parts. Reference signs not mentioned in the description shall not appear in the drawing, and vice versa.

(g) *Symbols, legends*. Graphical drawing symbols and other labeled representations may be used for conventional elements when appropriate, subject to approval by the Office. The elements for which such symbols and labeled representations are used must be adequately identified in the specification. While descriptive matter on drawings is not permitted, suitable legends may be used, or may be required, in proper cases, as in diagrammatic views and flowsheets or to show materials or where labeled representations are employed to illustrate conventional elements. Arrows may be required, in proper cases, to show direction of movement. The lettering should be as large as, or larger than, the reference characters.

(h) [Reserved]

(i) *Views*. The drawing must contain as many figures as may be necessary to show the invention; the figures should be consecutively numbered if possible in the order in which they appear. The figures may be plan, elevation, section, or perspective views, and detail views of portions or elements, on a larger scale if necessary, may also be used. Exploded view, with the separated parts of the same figure embraced by a bracket, to show the relationship or order of assembly of various parts are permissible. When an exploded view is shown in a figure which is on the same sheet as another figure, the exploded view should be placed in brackets. When necessary, a view of a large machine or device in its entirety may be broken and extended over several sheets if there is no loss in facility of understanding the view. Where figures on two or more sheets form in effect a single complete figure, the figures on the several sheets should be so arranged that the complete figure can be understood by laying the drawing sheets adjacent to one another. The figures, even though on separate sheets, should be labeled as separate figures, for example as Fig. 1a, Fig. 1b, etc., so that it would be apparent that views actually comprise one figure. The arrangement should be such that no part of any of the figures appearing on the various sheets is concealed and that the complete figure can be understood even though spaces will occur in the complete figure because of the margins on the drawing sheets. The plane upon which a sectional view is taken should be indicated on the general view by a broken line, the ends of which should be designated by numerals corresponding to the figure number of the sectional view and have arrows applied to indicate the direction in which the view is taken. A moved position may be shown by a broken line superimposed upon a suitable figure if this can be done without crowding, otherwise a separate figure must be used for this purpose. Modified forms of construction can only be shown in separate figures. Views should not be connected by projection lines nor should centerlines be used. When a portion of a figure is enlarged for magnification purposes, the figure and the enlarged figure must be labeled as separate figures.

(j) *Arrangement of views*. All views on the same sheet should stand in the same direction and, if possible, stand so that they can be read with the sheet held in an upright position. If views longer than the width of the sheet are necessary for the clearest illustration of the invention, the sheet may be turned on its side so that the top of the sheet with the appropriate top margin to be used as the heading space is on the right-hand side. One figure must not be placed upon another or within the outline of another.

(k) *Figure for Official Gazette*. The drawing should, as far as possible, be so planned that one of the views will be suitable for publication in the *Official Gazette* as the illustration of the invention.

(1) *Identification of drawings*. Identifying indicia (such as the application number, group art unit, title of the invention, attorney's docket number, inventor's name, number of sheets, etc.) not to exceed 2¾ inches (7.0 cm.) in width may be placed in a centered location between the side edges within three-fourths inch (19.1 mm.) of the top edge. Either this marking technique on the front of the drawing or the placement, although not preferred, of this information and the title of the invention on the back of the drawings is acceptable. Authorized security markings may be placed on the drawings provided they are outside the illustrations and are removed when the material is declassified. Other extraneous matter will not be permitted upon the face of a drawing.

(m) *Transmission of drawings*. Drawings transmitted to the Office should be sent flat, protected by a sheet of heavy binder's board, or may be rolled for transmission in a suitable mailing tube; but must never be folded. If received creased or mutilated, new drawings will be required.

(n) *Numbering of drawing sheets*. The drawing sheets may be numbered in consecutive arabic numbers at the top of the sheets, in the middle, but not in the margin. Such numbering will be deleted for printing purposes since page numbers are added at the time of printing the patent by the Office.

(o) *Copyright of Mask Work Notice*. A copyright or mask work notice may appear in the drawing but must be placed within the "sight" of the drawing immediately below the figure representing the copyright or mask work material and be limited to letters having a print size of 1/8 to ¼ inches (3.2 to 6.4 mm.) high. The content of the notice must be limited to only those elements required by law. For example, "© John Doe" (17 U.S.C. 401) and "*M* John Doe" (17 U.S.C. 909) would be properly limited and, under current statutes, legally sufficient notices of copyright and mask work, respectively. Inclusion of a copyright or mask work notice will be permitted only if the authorization language set forth in §1.71(e) is included at the beginning (preferably as the first paragraph) of the specification.

(p) *Limited use of color drawings in utility patent applications*. Paragraph (a) of this section requires that drawings in utility patent applications must be in black on white paper. However, on rare occasion, color drawings may be necessary as the only practical medium by which to disclose the subject matter sought to be patented in a utility patent application. The Patent and Trademark Office will accept color drawings in utility patent applications only after granting of a petition by the applicant under §1.183 of this part which requests waiver of the requirements of paragraph (a) of this section. Any such petition should be directed to the Office of the Deputy Assistant Commissioner for Patents and must include the following:

(1) The appropriate fee set forth in §1.17(h).

(2) Five (5) sets of color drawings on DIN size A4 (21.0 by 29.7 cm.) sheets.

(3) As proposed amendment to insert in the specification the following language as the specification relating to the brief description of the drawing:

> The file of this patent contains at least one drawing executed in color. Copies of this patent with color drawing(s) will be provided by the Patent and Trademark Office upon request and payment of the necessary fee.

(See §1.152 for design drawing, §1.165 for plant drawings, and §1.174 for reissue drawings.)

(Pub. L. 94-131, 89 Stat. 685)

[24 FR 10332, Dec. 22, 1959, as amended at 31 FR 12923, Oct. 4, 1966; 36 FR 9775, May 28, 1971; 43 FR 20464, May 11, 1978; 45 FR 73657, Nov. 6, 1980; 53 FR 47809, Nov. 28, 1988]

Appendix 17
Comparison of Classification Schedules from 1923 (Part A) to 1986 (Part B)

PART A

DEPARTMENT OF THE INTERIOR
Hubert Work, Secretary

UNITED STATES PATENT OFFICE
Thomas E. Robertson, Commissioner

MANUAL OF

CLASSIFICATION OF PATENTS

"Definitions of Revised Classes," published in 1912, and "Classification Bulletins," numbered consecutively beginning with No. 28 and published semiannually, are referred to in the body of the book in connection with each class defined.

Substitute pages will be published from time to time as changes are made, and may be procured from

COMMISSIONER OF PATENTS
WASHINGTON, D. C.

An INDEX will be printed separately

WASHINGTON
GOVERNMENT PRINTING OFFICE
1923

Page 97

111. PLANTING—Drilling—Hill-planting machines—Depositing mechanisms—Continued.

42 Endless-belt tripped,
43 Line-wire driven
44 Reel carried,
45 Check-wire tripped
46 Both feeder and valve,
47 Trip-fork mechanisms,
48 Guides,
49 Wire-end-anchoring devices,
50 Manually operated.
51 Accumulators,
52 Frame and planting-element arrangement
53 Sectional main frame
54 Flexible
55 With auxiliary frame,
56 Break joint,
57 Extensible
58 V-shaped,
59 Main and auxiliary frame
60 Plurality of auxiliaries
61 Unitarily controlled,
62 Floating auxiliary
63 Hopper carrying
64 Detachable
65 Single row,
66 Tool-bar type
67 With lift and ungear,
68 With crank-axle lift,
69 With adjustable planter elements,
70 Rigid
71 Single row
72 Manually operated depositing mechanism,
73 Multiple depositing,

111. PLANTING—Drilling—Frame and planting-element arrangement—Rigid—Single row—Continued.

74 Revolving hopper,
75 Vibrating hopper,
76 Vibrating delivery chute.
77 Rotating dispenser
78 Axle mounted,
79 Single-row implements
80 Multiple depositing,
81 Planter-element arrangement,
82 Hand propelled,
83 Lister units,
84 Drag-bar units,
85 Drill sets,
86 Drill teeth
87 Rotary furrower
88 Multiple disk.
89 Dibbling
90 Revolving-hopper implements
91 Revolving-dibble-carrier implements,
92 Manually operated implements
93 Machine attached,
94 Spacing in hill,
95 Regulated discharge
96 Sliding-plunger control,
97 Multiple-staff control,
98 Footplate control,

112 SEWING MACHINES.
Definitions in Bulletin 42.

1 Miscellaneous.
2 Special machines
3 Mattress sewing,
4 Jacquard-card sewing,
5 Lacing shoe-uppers,
6 Broom-sewing,

112. SEWING MACHINES—Special machines—Continued.

7 Carpet sewing
8 Clamps,
9 Rug sewing,
10 Bag sewing,
11 Sewing filled sacks,
12 Sewing hats
13 Work supports
14 Rotary,
15 Guides and gages,
16 Glove and fur sewing
17 Dyeing feature,
18 Work holding and feeding
19 Clamp,
20 Guiding,
21 Book sewing
22 Knotting,
23 Straw braid,
24 Straw sewing,
25 Looped fabrics
26 Preparing,
27 Work holding and feeding.
28 Leather sewing
Shoe sole
29 Separating and indenting,
30 Tack pulling,
31 Fair stitch,
32 Loop lock,
33 Tongue lock,
34 Chain stitch
35 Curved needle,
36 Lock stitch
37 Curved needle
38 Oscillating and rotating shuttle,
39 Back gages and rests,
40 Guards,
41 Heating,

112. SEWING MACHINES—Special machines—Leather sewing—Continued.

42 Waxing,
43 Lubricating and moistening,
44 Welt slitting and beveling,
45 Channeling,
46 Welt handling,
47 Feeding
48 Awl,
49 Needle,
50 Channel guides,
51 Edge and crease guides,
52 Welt guides,
53 Chain stitch,
54 Lock stitch,
Thread handling
55 Orbital-movement loopers,
56 Cast-offs,
57 Take-ups,
58 Pull-offs,
59 Tensions,
60 Presser devices
61 Locking,
62 Work supports,
63 Tube forming,
64 Fringe forming,
65 Buttonhole
66 Eyelet making,
67 Start and stop mechanism,
68 Cutting,
69 Thrum mechanism,
70 Clamp feeding
71 With rotatable needle bar,
72 Right line and rotary,
73 Vibrating needle,
74 Spreading,
75 Gaging,
76 Clamps,
77 Attachments.

112. SEWING MACHINES—Special machines—Continued.

78 Embroidering
79 Turfing
80 Hand implements,
81 Hemstitch
82 Universal feed,
83 Horizontal needle
84 Pattern controlled
85 Boring,
86 Fabric shifting,
87 Stopping,
88 Article attaching,
89 Boring,
90 Fabric shifting
91 Balancing
92 Spring,
93 Stitch forming
94 Short thread,
95 Shuttle type,
96 Thread take-up,
97 Thread tension,
98 Vertical needle
99 Attaching ornaments
100 Thread
101 Attachments,
102 Fabric shifting
103 Frames,
104 Article attaching
105 Hook and eye
106 Feeding,
107 Carriers,
108 Eye-shank button
109 Vibrating needle,
110 Flat button
111 Vibrated needle,
112 Vibrated clamp,
113 Feeding,
114 Clamps,
115 Attachments,

112. SEWING MACHINES—Special machines—Continued.

116 Crocheting,
117 Quilting
118 Fabric shifting
119 Frames,
120 Infolding,
121 Darning,
Work manipulating
122 Trimmers
123 With stitching mechanism,
124 Rotary,
125 Manual control,
126 Laterally adjustable,
127 Ply separating,
128 Needle-bar operated,
129 Cutters,
130 Severing,
131 Creasers and markers,
132 Rufflers and gatherers
133 Smocking,
134 Attachments
135 Vibrating blade,
136 Guides
137 Binders
138 Strip reversing,
139 Braiders and corders,
140 Blind stitch,
141 Hemmers
142 Lap seam,
143 Adjustable,
144 Tucking
145 Box plait,
146 Spacing,
147 Folding,
148 Fabric holding,
149 Fringe turning,
150 Pressure,
151 Presser bar attached,

112. SEWING MACHINES—Work manipulating—Guides—Continued.

152 Strip,
153 Edge,
154 Stitch forming
155 Multiple machines,
156 Knot tying,
157 Zigzag
158 Vibrating needle
159 Shifting looper,
160 Surging,
161 Purling,
162 Overseaming,
163 Multiple needle
164 Shuttle type,
165 Chain and double chain
166 Multiple looper,
167 Adjustable,
168 Convertible,
169 Hand implements,
170 Short thread
171 Double-pointed needle,
172 Overedge,
173 Running stitch
174 Crimping,
175 Stationary needle,
176 Blind stitch
177 Overseaming,
178 Work handling,
179 False hemstitch,
180 Bobbin ejector,
181 Revolving-hook type
182 Timing,
183 Reciprocating bobbin,
184 Vertical axis,
185 Shuttle type
186 Shuttle replenishing,
187 Loop spreaders,
188 Bobbin latch,

112. SEWING MACHINES—Stitch forming—Shuttle type—Continued.

189 Rotary
190 Timing,
191 Vertical axis,
192 Oscillatory
193 Vertical axis,
194 Reciprocatory
195 Longitudinal,
196 Holders and raceways,
197 Chain or double chain
198 Hooked needle,
199 Oscillating or reciprocating looper,
200 Four-motion looper,
201 Rotating looper,
202 Vertical-axis looper,
203 Feeding
204 Universal,
205 Irregular,
206 Needle and helper,
207 Clamp,
208 Differential
209 Adjustment,
210 Adjustment
211 Rotary,
212 Upper
213 Needle,
214 Rotary,
215 Four motion,
216 Dogs,
217 Seam finishing,
218 **Elements**
219 Starting and stopping,
220 Driving mechanisms
221 Needle,
222 Needles
223 Thread releasing,
224 Threading structure,

PART B

112-1

CLASS 112 SEWING

DECEMBER 1986

400	PRODUCTS
401	.Uniplanar sectional web, sheet or layer
402	.Sewn web or sheet
403	..Including decomposable thread or component
404	..Including feathers, beads or reflective material
405	..With longitudinally non-coextensive super-imposed panel or sheet
406	..Including external fastener sewn thereto
407	...Hook and/or eye
408	...Button
409	..Fringed
410	..Tufted or looped surface
411	...With tufts or loops formed by floating strand-like portions
412	..Of strands or strand portions joined by sewing
413	..Elastic strand or component
414	...Including non-planar component
415	..Specific strand arrangement or relationship
416	...Knitted, netted or knotted
417	..With strip, strand or strand portion between layers or components
418	..With strip sewn in or over seam or over edge (e.g., welt or stay)
419	...Channel-shaped edge binding
420	..Fibrous or particulate layer between other layers
421	..Including component of varying thickness
422	..Non-planar uniform-thickness material
423	...Bent flange or reverse fold at edge
424	....Including concealed or blind stitching
425	.....Overedge or along offset fold
426	....Included in seam joining contiguous components
427	...Pleated, tucked or shirred
428	...Plural components
429	..Filamentary material or strand-like tape sewn to surface
430	...Interlooped floating strands or strand portions
431	...Around aperture (e.g., buttonhole)
432	...Connecting parallel straight lines of stitching
433	....Included in overedge stitching
434	....Uniting contiguous components
435	.....Plural intersecting float threads
436	...Incorporated with or held by overedge stitching
437	..Apertured or open-work (e.g., buttonhole)
438	..Chain stitch
439	..Ornamental stitching (e.g., embroidery)
440	..United components
441	...By stitching along edge
2	SPECIAL MACHINES
2.1	.Mattress sewing
2.2	..Tufting
4	.Jacquard card sewing
6	.Broom sewing
7	.Carpet sewing
8	..Clamps
9	.Rug sewing
10	.Bag sewing
11	.Sewing filled sacks
12	.Sewing hats
13	..Work supports
14	...Rotary
15	..Guides and pages
16	.Glove and fur sewing
17	..Dyeing feature
18	..Work holding and feeding
19	...Clamp
20	..Guiding
21	.Book sewing
22	..Knotting
23	.Straw braid
24	.Straw sewing
25	.Looped fabrics
26	..Preparing
27	..Work holding and feeding
28	.Leather sewing
	..Shoe sole
29	...Separating and indenting
30	...Tack pulling
31	...Fair stitch
32	...Loop lock
33	...Tongue lock
34	...Chain stitch
35	....Curved needle
36	...Lock stitch
37	....Curved needle
38	.....Oscillating and rotating shuttle
39	...Back gauges and rests
40	...Guards
41	..Heating
42	..Waxing
43	..Lubricating and moistening
44	..Welt slitting and beveling
45	..Channeling
46	..Welt handling
47	..Feeding
48	...Awl
49	...Needle
50	..Channel guides
51	..Edge and crease guides
52	..Welt guides
53	..Chain stitch
54	..Lock stitch
	..Thread handling
55	...Orbital-movement loopers
56	...Cast-offs
57	...Take-ups
58	...Pull-offs
59	...Tensions
60	..Presser devices
61	...Locking
62	..Work supports
63	.Tube forming
64	.Fringe forming
65	.Buttonhole
66	..Eyelet making
67	..Start and stop mechanism
68	..Cutting
69	..Thrum mechanism
70	..Clamp feeding
71	...With rotatable needle bar
72	...Right line and rotary
73	...Vibrating needle
74	...Spreading
75	...Gauging
76	...Clamps
77	...Attachments
78	.Embroidering
80.01	..Tufting (replacing subclass 79)
80.02	...Rooting hair in doll or wig
80.03	...Hand implement
80.04	....With power drive
80.05	....Having hollow or channelled needle
80.06	....Having non-eyed needle
80.07	...Yarn manipulation by fluid flow
80.08	....Having hollow needle
80.15	...Dual sided
80.16	...Including hollow or channelled needle
80.17	...Including non-eyed needle
80.18	...With condition responsive stop motion means
80.23	...With optical, electronic, or magnetic pattern program means
80.24	....Having pattern drum

	SPECIAL MACHINES
	.Embroidering
	..Tufting (replacing subclass 79)
80.3	...With specific fabric supporting, manipulating, cutting, or treating means
80.31	....Means to shift fabric laterally of feed
80.32	....Means to feed fabric
80.33	....Fabric vertically moveable at tufting position
80.4	...Including specific needle supporting or manipulating means
80.41	....Means to shift needle laterally of fabric feed
80.42	....Means to vertically adjust stroke
80.43	....Means to selectively drive one of plural needles
80.44	....Means to disconnect needle and drive
80.45	....Supporting structure
80.5	...Including specific loop catcher
80.51	....Bill with clip or gate
80.52	....Plural loop catchers having different shapes or orientations
80.53	.....Stationary relative to one another
80.54	....Variable height loops
80.55	....With loop cutting means
80.56	.....Selective cutting means
80.57	.....Rotary cutter
80.58	.....Knife pivots on loop catcher
80.59	.....Cutting means stationary relative to loop catcher
80.6	.....Including supporting structure (e.g., knife block)
80.7	...Including specific yarn manipulating, cutting, or treating means
80.71	....Cutting or treating means
80.72	....Feed via intermeshing slats
80.73	....Feed roller
81	..Hemstitch
82	...Universal feed
83	..Horizontal needle
84	...Pattern controlled
85	....Boring
86	....Fabric shifting
87	...Stopping
88	...Article attaching
89	...Boring
90	...Fabric shifting
91	....Balancing
92	....Spring
93	...Stitch forming
94	....Short thread
95	....Shuttle type
96	....Thread take-up
97	....Thread tension
98	..Vertical needle
99	...Attaching ornaments
100	....Thread
101	.....Attachments
102	...Fabric shifting
103	....Frames
104	.Article attaching
105	..Hook and eye
106	...Feeding
107	...Carriers
108	..Eye-shank button
109	...Vibrating needle
110	..Flat button
111	...Vibrated needle
112	...Vibrated clamp
113	..Feeding
114	..Clamps
115	..Attachments
116	.Crocheting
117	.Quilting
118	..Fabric shifting
119	...Frames
120	.Infolding
121	.Darning
121.11	.Pattern controlled or programmed
121.12	..Work carrier and support
121.13	.Stitch control
121.14	.Traveling
121.15	.Work carrier and support
121.16	.Tuft making
121.17	.Feather sewing
121.18	.Spiral feed
121.19	.Stringing
121.2	.Lacing or whipping, e.g., covering rings
121.21	.Sewing hemp-sole shoes
121.22	.Neckties
121.23	.Upholstery
121.24	.Circular line of stitch
121.25	.Stitch marking
121.26	.Fabric tensioning
121.27	.Elongated articles, e.g., belt, waistband and belt loop
121.28	.Spherical object, e.g., baseball cover sewing
121.29	.Completed-work handling
	WORK MANIPULATING
122	.Trimmers
122.1	..Having specified stitching mechanism
122.2	...Pinker
122.3	..Rotary cutter
122.4	...Pinker
125	..Manual control
126	..Laterally adjustable
127	..Ply separating
128	..Needle-bar operated
129	..Cutters
130	.Severing
131	.Creasers and markers
132	.Rufflers and gatherers
133	..Smocking
134	..Attachments
135	...Vibrating blade
136	.Guides
137	..Binders
138	...Strip reversing
139	..Braiders and corders
140	..Blind stitch
141	..Hemmers
142	...Lap seam
143	...Adjustable
144	..Tucking
145	...Box plait
146	...Spacing
147	..Folding
148	..Fabric holding
149	..Fringe turning
150	..Pressure
151	..Presser bar attached
152	..Strip
153	..Edge
154	STITCH FORMING
155	.Multiple machines
156	.Knot tying
157	.Zigzag
443	..Vibrating needle
159	...Shifting looper
444	...Having indicator
445	....And specified electronic memory
446	...Closed pattern sewing (e.g., buttonhole)
447	....Including electronic memory
448	....Having cam actuation of needle or material feed
449	.....And means to change, without operator intervention, stitch width or material feed
450	...Basting stitch or skipping stitch
451	...Backstitch
452	...Having multiple needles

Appendix 18
Classification Definitions

U.S. PATENT AND TRADEMARK OFFICE

DOCUMENTATION ORGANIZATIONS

PATENT CLASSIFICATION DEFINITIONS

PLANTS

DECEMBER 1987

PLANTS

CLASS DEFINITIONS

I. This is the class for plants which are patentable under 35 U.S. Code, Section 161, which provides for the granting of a patent to whoever invents or discovers and asexually reproduces any distinct and new variety of plant, including cultivated sports, mutants, hybrids, and newly found seedlings other than a tuber propagated plant or a plant found in an uncultivated state.

II. The scheme of classification is that a patent or publication is placed in the first appearing of a series of co-ordinate subclasses which the subject matter of the patent or publication fits. Thus, a patent describing a plant which is both a conifer and a shrub is classified as a conifer in subclass 50 rather than a shrub in subclass 54. A patent not fitting the description of any of the first line subclasses is classified in miscellaneous subclass 89.

A subclass which is positioned one space to the left of one or more following subclasses is considered to be the miscellaneous subclass for that group of subclasses. Thus, subclass 54 from Shrubs or Vines takes those shrubs or vines which are not azaleas or rhododendrons, barberries, buddleias, etc., while subclass 60 for Camellias takes those camellias which are neither pink nor red.

III. The color definitions given herein are substantially those found in Webster's New International Dictionary, Second Ed., Unabridged, published by G. C. Merriam Co., Springfield, Mass. Each color definition refers to one or more plant patents so that the blossoms of the plant patents themselves comprise a kind of color chart to illustrate the words of the definitions.

The color designations apply to the color of the blossom when it is newly open and in an unfaded condition, i.e., not in bud or full-blown. The color actually shown in the drawing is the color used to determine placement of patents, except in those cases where the description in the specification indicates the color of the drawing is not accurate, in which case the patent is placed in accordance with the total disclosure furnished in the patent. In the various examples of plant patents given in the definitions the color of the drawing alone is the color referred to.

The effects of light and shadow should be discounted when determining the true color of a blossom for purposes of classification. Also, the color at the base of the petal should be disregarded except where a two-tone or bicolor effect is quite obvious.

All the color designations refer to solid colors unless clearly indicated otherwise. Thus, considering the group of climbing roses, a striped or bicolor rose would not fit any of the indented subclasses but would be placed in the miscellaneous subclass for climbing roses, subclass 2. In determining whether or not a blossom has a solid color the appearance of the flower as a whole is the proper criterion. Minor flecks and gradations of color should be disregarded. However, both faces of all petals must be substantially the same color.

SEARCH CLASS:

800, Multicellular Living Organisms and Unmodified Parts Thereof, appropriate subclasses for living multicellular organism, e.g., plants, etc., and separated or severed parts thereof that have not undergone any modification or treatment subsequent to their separation, e.g., untreated seeds, etc.

SUBCLASS DEFINITIONS

Definitions and notes are given only for those subclasses whose titles are considered not completely self-explanatory.

1. Rose plants under the class definition.

(1) Note. This subclass is the home for "shrub" roses such as species hybirds, hybird perpetuals, etc., characterized by limited flowering, coarseness of the blooms and/or plant and superior cold-hardiness.

2. Rose plants under subclass 1 and characterized by vigorous, long, erect or lax canes suitable for training on trellises or fences or suitable as ground covers.

    (1) Note. Pillar and Rambler roses are included under this definition.

3. Climbing roses under subclass 2 characterized by blossoms which are white (a color comparable to fresh snow; a neutral or achromatic color of highest brilliance; the lightest gray), as typified by plant patents, Nos. 267, 400 and 1273 or yellow (a color which resembles the hue of ripe lemons; the color of sulphur), as typified by plant patents, Nos. 704, 1263 and 1557.

4. Climbing roses under subclass 2 characterized by blossoms which are orange (a color varying from reddish red-yellow to red yellow, in saturation from high to very high and in brilliance from medium to high), as typified by plant patents, Nos. 106, 458 and 2072 or salmon (a color which is reddish red-yellow, of medium saturation and high brilliance, as typified by plant patent, No. 1606.

5. Climbing roses under subclass 2 characterized by blossoms which are pink (a color varying from reddish blue-red to yellowish-red, from low to medium saturation and from high to very high brilliance), as typified by plant patents, Nos. 287, 385 and 893.

6. Climbing roses under subclass 2 characterized by blossoms which are red (a color ranging from that of blood to that of the ruby), as typified by plant patents, Nos. 856, 1366 and 1864.

7. Rose plants under subclass 1 which range in height from a few inches to approximately 1 foot and which have blossoms less than 1 inch in diameter.

8. Miniature roses under subclass 7 characterized by blossoms which are white (a color comparable to fresh snow; a neutral or achromatic color of highest brilliance; the lightest gray), as typified by plant patents, Nos. 1407, 1412 and 1908 or yellow (a color which resembles the hue of ripe lemons; the color of sulphur), as typified by plant patents, Nos. 407, 1631 and 2177.

9. Miniature roses under subclass 7 characterized by blossoms which are pink (a color varying from reddish blue-red to yellowish-red, from low to medium saturation and from high to very high brilliance), as typified by plant patents, Nos. 1293, 1451 and 1984.

10. Miniature roses under subclass 7 characterized by blossoms which are red (a color ranging from that of blood to that of the ruby), as typified by plant patents, Nos. 1032, 1663 and 2080.

11. Rose plants under subclass 1 characterized as free-flowering with large, well-shaped blooms borne singly or in clusters on long stems.

12. Hybrid tea or grandiflora roses under subclass 11 characterized by blooms in which either the reverse or face sides of the petals are red (a color ranging from that of blood to that of the ruby), as typified by plant patents, Nos. 378, 959 and 1489 and the other sides of the petals are a color other than red.

13. Hybrid tea or grandiflora roses under subclass 11 characterized by blooms exhibiting a splashing, striping, speckling or dotting of two or more distinct colors, as typified by plant patents, Nos. 167, 197 and 1582.

14. Hybrid tea or grandiflora roses under subclass 11 characterized by blooms which are white (a color comparable to fresh snow; a neutral or achromatic color of highest brilliance; the lightest gray), as typified by plant patents, Nos. 1530, 1762 and 1770.

15. Hybrid tea or grandiflora roses under subclass 11 characterized by blooms which are yellow (a color which resembles the hue of ripe lemons; the color of sulphur), as typified by plant patents, Nos. 753, 1643 and 2089.

Appendix 19
Class Schedules from *Manual of Classification*

MANUAL
OF
CLASSIFICATION

BASIC MANUAL
DECEMBER 1988

Revision No. 3

Including changes resulting from
all Classification Orders through

June 30, 1990

U. S. DEPARTMENT OF COMMERCE
PATENT AND TRADEMARK OFFICE
DOCUMENTATION ORGANIZATIONS

The Patent and Trademark Office does not handle the sale of this Manual, distribution of notices and revisions or changes of address of those on the subscription list.

Correspondence relating to any of the above items should be sent to the Superintendent of Documents at the following address:

Superintendent of Documents
Mail List Section
Washington, D. C. 20402

For Sale by the Superintendent of Documents, U. S. Government Printing Office
Washington, D. C. 20402

CLASS 172 EARTH WORKING

DECEMBER 1988

1 PROCESSES
2 AUTOMATIC POWER CONTROL
3 .Motive power control
4 .Constant depth type
4.5 .Land leveller type
5 .Obstruction sensing type (includes plant sensing)
6 ..Electrical
7 .Draft responsive
8 ..Variable rate responsive
9 ..With manual actuator to select type of condition sensed
10 ..Sensitivity adjustment
11 ..With excess draft release
12 ..Overload lift type
13 LAWN EDGER
14 .With or convertible to non-earth working implement
15 .Rolling or driven cutter
16 ..With fixed cutter or furrower
17 .With wheel or roller
18 .Impact or grapple
19 SOD CUTTER
20 .With means for vertical transverse cutting while moving
21 LAWN AERATOR OR PERFORATOR, OR PLUG REMOVER
22 .Earth removing
23 DRIVEN FROM OR GUIDED BY STATIONARY OBJECT, OR ANCHORED
24 .Around tree or stake
25 .Rotatable about vertical axis
26 .Guided by surface track or previously formed shoulder
26.5 .Dragline scraper
26.6 ..Scraper part rearranged upon reverse movement
27 WITH MEANS FOR CUTTING OR SHREDDING PLANTS WITHOUT SOIL DISTURBANCE
28 .Driven
29 WITH MEANS FOR SHIFTING SURFACE MATERIAL WITHOUT SOIL DISTURBANCE
30 .Driven shifting means
31 .Combined with rolling or vertically acting transverse cutter
32 WITH SEPARATING AFTER EARTH WORKING
33 WITH POWER DRIVEN MOLDBOARD, CONVEYER OR HANDLER
34 COMPLETE APPARATUS ADAPTED FOR USE UPSIDE DOWN
35 WITH DRIVE MEANS FOR TOOL OR CLEANER
36 .Subsurface shears or nippers
37 .Tool rotated by attendant
38 .With obstruction feeling device for moving or releasing implement
39 .With cleaner or comminutor spaced from ground surface
40 .Vibrating tool
41 .Attendant supported tool
42 .Guided by walking attendant
43 ..With ground support vertically adjustable relative to frame
44 .Subsurface shaft or bar (e.g., rod weeder)
45 .Flails
46 .Coaxial tools oppositely rotated
47 .With specific relationship of mast-type hitch (i.e., three-point hitch) to implement
48 .Plural driven tools
49 ..Contiguous cooperating or intermeshing rotary ground engaging tools
49.5 ...Rotating about vertical axes
50 ..Diverse tools
51 ...All rotary
52 ....Parallel axes
53 ...Rectilinearly reciprocating tool
54 ...Oscillating tool
54.5 ..Tool reciprocates or oscillates within a generally horizontal plane
55 ..Plural groups of disks
56 ..Staggered tools
57 ..Laterally spaced tools
58 ...Longitudinal axes
59 ...Vertical axes
60 ...Transverse axes
61 .Intermittent drive for tool
62 ..With spring return
63 .With non-driven tool (e.g., plow, harrow, drag, scraper, knife or roll, etc.)
64 ..Non-driven furrow opener and driven dam former
65 ..Interdigitating non-driven and driven tools
66 ..Cooperating driven cleaner or comminutor and contiguous tool
67 ..Driven comminutor at outlet of earth guide
68 ..Rolling tool
69 ...With tool drive from rolling tool
70 ..Fore-and-aft non-driven tool
71 ..Non-driven tool follows path of driven tool
72 ...Leveling drag or furrow shaper
73 ..Staggered driven and non-driven tool (e.g., cotton chopper, etc.)
74 .With power take-off from tool drive to adjust tool
75 .Interconnected tool lift and drive control
76 .Implement with ground support for depth control
77 ..Vertically biased implement
78 ..Vertically adjustable ground support
79 ..Tool driven from prime mover on vehicle
80 .With wheel substitute (e.g., runner, etc.)
81 .With plant deflector or protector
82 .Driven tool selectively shiftable along line of travel
83 ..Tool drive interrupted by shifting tool
84 .Simultaneously reciprocating and oscillating blade having elongated shank
85 ..Transverse chopping type
86 ...With plural cranks or cams driving each blade
87 ...Means for varying contour of path of blade
88 ..With plural cranks or cams driving each blade
89 ..Means for varying contour of path of blade
90 .Irregular or off-center ground-engaging wheel or support
91 .Blade movable with respect to cyclically driven carrier
92 ..With means for moving blade
93 ...Rectilinearly reciprocating blade
94 ...Blade oscillating arcuately or swivelly with respect to rotary carrier
95 ...By cam or crank
96 ..Blade flexible or with yieldable mount on carrier
97 .Compound motion for tool (e.g., reciprocating and oscillating, reciprocating and rotating)
98 .Tool mounted for lateral shifting
99 ..About generally vertical axis
100 .Blade on endless driven belt or chain
101 .Tool guided for rectilinear reciprocation

172-2

CLASS 172 EARTH WORKING

DECEMBER 1988

	WITH DRIVE MEANS FOR TOOL OR CLEANER
	.Tool guided for rectilinear reciprocation
102	..Tool moves in horizontal, transverse path
103	.With overload relief or clutch in drive train (e.g., overload release, etc.)
104	..Unidirectional clutch in drive from ground wheel
105	.Driven from rolling or driven ground wheel
106	..Belt or chain drive
107	.Tool driven about horizontal, longitudinal axis
108	..Rotary driven tool
109	...Adjustable tooth or blade
110	.Tool driven about generally vertical axis (e.g., oscillating choppers, etc.)
111	..Rotary driven tool
112	.With deflector or shield for thrown material
113	..Laterally directed outlet flow
114	.Specific propelling means
115	..Tool steers implement
116	..Tool propels implement
117	.Tool freely or yieldably mounted on chassis
118	.Tool driven about axis transverse to draft line
119	..Screw or spiral rib, blade or tooth row
120	..Disk or planar cutter (e.g., saw, etc.)
121	..Laterally extending bar or blade with skeleton support (e.g., lawn mower type, etc.)
122	..Drum with teeth or blades
123	..Rotary driven tool
124	.Tool driven about diagonal axis
125	.Tool drive details
126	WITH EARTH MARKER
127	.Marker shiftable on turning
128	.Marker adjusted upon raising implement
129	.Ground wheel operated marker control
130	.Multiple interconnected markers
131	..Markers on laterally shiftable member
132	.Marker swingable about longitudinal axis to both sides
777	SCRAPER SUPPORTS NARROW DEPENDING TOOL
778	.Tool supporting clamp means engage upper and lower edges
779	SCRAPER POSITION AUTOMATICALLY CONTROLLED BY LINKAGE FOR LEVELLING
780	SCRAPER BETWEEN WIDELY SPACED FRONT AND REAR GROUND SUPPORTS
781	SCRAPER BETWEEN FRONT AND REAR GROUND SUPPORTS OF VEHICLE
782	.With laterally offset inclined shoulder forming tool
783	.With scraper attached ground support
784	.With diverse tool or portion
785	..Non-scraping tool precedes and spaced from scraper
786	.Plural scrapers
787	..Spaced and in same path
788	.Push frame for scrapers
789	.Actuator for bodily shifting scraper subframe draft connection
790	.Counterbalance means for scraper adjustment
791	.Three or more independently operable scraper actuators
792	..Scraper adjustable about vertical axis of annular support
793	...Actuator for laterally shifting support
794	.Spring biased into ground contact
795	.Specific actuator between frame and scraper
796	..For adjustment about vertical axis
797	..For adjustment about longitudinal axis
798	.Actuator for tilting wheel relative to vehicle frame
799	.Specific means for horizontally angling wheel relative to vehicle frame
799.5	TOWED SCRAPER WITH GROUND SUPPORT WHEELS
133	DIVERSE TOOLS
134	.One located in path of implement wheel
135	.One implement surrounds another
136	.Tools usable alternately only
137	.With means to vary spacing of tools upon turning
138	.With interconnected vertical adjustment
139	..Plow and colter
140	.With independent means for vertical movement
141	.Interconnected adjustment of horizontal angle of rolling and position of diverse tool
142	.Including spring formed tool or standard
143	.Including intermittently rolling tool
144	.Colter, jointer and plow
145	.Three or more diverse implements following same path (A, B, C, or A, B, A,)
146	..Four or more
147	...Alternately diverse (A, B, A, B)
148	..Longitudinally spaced like implements with intermediate diverse implement (A, B, A)
149	..Including rolling tool
150	...Smooth levelling roller
151	...Diverse rolling
152	.At least four alternately diverse laterally spaced tools (A, B, A, B)
153	..Alternate rolling and non-rolling
154	..All rolling
155	.Laterally spaced like tools with intermediate diverse tool (A, B, A)
156	..Spaced rolling with intermediate nonrolling
157	..Spaced non-rolling with intermediate rolling
158	..All rolling
159	..Spaced right and left hand tools with intermediate symmetrical tool
160	..Including spike tooth
161	.Including implement alternating for right or left hand operation
162	..Reversal of implement adjusts diverse tool
163	.Jointer and plow
164	..Rolling jointer
165	.Including colter
166	..Rolling colter
167	.Fixed point or share with rotary moldboard
168	.Rotating tool with fixed moldboard
169	.Including tool rotatable about vertical axis
170	.Including smooth levelling roller
171	..Spaced from moldboard side of plow
172	..With diverse rolling tool
173	..With teeth
174	.Rolling and non-rolling
175	..Following same path
176	...Furrowing or ridging implement followed by furrow or ridge roller
177	...Rolling tool has circumferentially spaced blades, tines or the like
178	...Including disk gang
179	...Non-rolling tool group with laterally co-extensive rolling tool

Subclass	Title
	AERIAL PROJECTILE, TARGET THEREFOR, OR ACCESSORY
	.Target
398	..Pocketed or apertured
399	...With mechanical projector
400	...Closed-back or closed-bottom pocket
401	....Entrance opening is formed in laterally extending surface
402	...Target is aperture dimensioned to allow projectile to pass entirely therethrough
403	..Penetrable target with replaceable element
404	..Penetrable target with projectile backstop
405	..With mechanical projector
406	..Handling or manipulation (e.g., target positioning means)
407	..Target support structure
408	..Target penetrated by projectile
409	..Printed matter
410	.Projectile backstop
411	.Playing field or court game; or accessory therefor (e.g., volleyball, soccer, etc.)
412	.User manipulated means for catching projectile moving through the air
413	.Tethered projectile
414	..Tethered to means engageable with human body part (e.g., hand-held means or means attachable to human body part)
415	.Scattershot projectile or beanbag projectile
416	.Arrow, dart, shuttlecock, or element thereof
417	..Shuttlecock
418	..Material dispensing upon impact or having fluid conducting means
419	..Head structure
420	...And vane structure (i.e., flight guiding or stabilizing means)
421	...Broadhead
422	....Interchangeable blade
423	..Vane structure (i.e., flight guiding or stabilizing means)
424	.Disc or ring projectile
425	..Ring
426	.Boomerang projectile
427	.Horseshoe projectile
428	.Nonspherical projectile
108	SURFACE PROJECTILE
109	.Moving surface
110	..Pivoted
111	..Pivoted gate
112	..Spiral surface
113	..Pocketed
114	...Mercury globule
115	...Surface pockets
116	....Hazard pockets
117	...Moving pockets
118 R	.Ball games
118 A	...Magnetic or electric
118 D	...Ball actuated element
119 R	..Combined with projector
119 A	....Electric or magnetic
119 B	....Fluid projector
120 R	...Gravity projectors
120 A	.....Electric or magnetic
121 R	...Return course
121 A	.....Electric or magnetic
121 B	.....Pachinko, i.e., nearly vertical playing surface
121 D	.....Ball supply means
121 E	.....Ball elevator
122 R	...Ball return
122 A	.....Electric or magnetic
123 R	..Pocketed
123 A	....Electric or magnetic
124 R	...Return course
124 A	.....Electric or magnetic
125 R	...Ball return
125 A	.....Electric or magnetic
126 R	.Disk or ring games
126 A	...Electric or magnetic
127 R	.Targets
127 A	...Fall apart targets
127 B	...Static, no ball return
127 C	...Ball return
127 D	...Target pivots about horizontal axis
128 R	.Projectiles
128 CS	...Curling stones
128 A	...Rollable
129 R	.Projectors
129 K	...Swung, carried by user
129 L	...Slid, carried by user
129 M	...Wheeled, carried by user
129 P	...Pinched or strand engaged
129 Q	...Gravity propelled
129 AP	...Air propelled
129 S	...Plunger, Mechanically or Electrically driven
129 T	...Plunger, Manual
129 V	...Pivoted, Mechanically or Electrically driven
129 W	...Pivoted, Manual
236	BOARD GAMES, PIECES OR BOARDS THEREFOR
237	.Electrical
238	..Removable and discrete game piece changes status of circuit
239	.Magnetic
240	.Markable or erasable game board or piece (e.g., magic slate)
241	.Having three-dimensional pattern
242	.Piece moves over board having pattern
243	..Chance device controls amount or directionof movement of piece
244	...Sports or outdoor recreational activities
245	....Golf
246	....Racing
247	....Football or soccer
248	...Race to a finish (e.g., backgammon)
249	....With common finish (e.g., parchisi)
250	.....Outer space or astronomy
251	.....Travel or exploration
252	....Travel or exploration
253	...Outer space or astronomy
254	...Travel or exploration (e.g., touring, treasure hunt, archeology)
255	...Military or naval engagement
256	...Property or commodity transactions
257	...Judicial, legislative, or election process
258	..Strategic race to a finish (e.g., Chinese checkers)
259	..Sports or outdoor recreational activities
260	..Chess or checker type
261	...Nonrectangular or extended pattern
262	..Military or naval engagement
263	..Chase type (e.g., fox and geese)
264	..Alignment games (e.g., morris, mill)
265	.Salvo type
266	.Nim type (i.e., game of take away)
267	.Completing square type
268	.Dice board and number plate type
269	.Lotto or bingo type
270	..With attached pieces
271	.Alignment games (e.g., tic-tac-toe, go-moko)
272	.Word, sentence, or equation forming (e.g.,SCRABBLE, hangman)
273	.Memory or matching games (e.g., concentrat

273-6

CLASS 273 AMUSEMENT DEVICES, GAMES

DECEMBER 1988

	BOARD GAMES, PIECES OR BOARDS THEREFOR
274	.Betting or wagering board (e.g., casino)
275	.Path forming
276	.Construction or assembly games
277	.Sports or outdoor recreational activities
278	.Property or commodity transaction (e.g., stock market)
279	.Judicial, legislative, or election process
280	.Rotatably mounted board
281	.Game board having movably attached piece
282	.Removably interfitting or detachably adhesive board and piece
283	.Game board having pattern separable into sections
284	.Game board having interchangeable, variable, or plural distinct playing patterns
285	.Collapsible board (e.g., folding)
286	..Flexible sheet type
287	.Game board structure
288	.Game piece
289	..With movably attached part
290	..Stackable or nestable feature
291	..Weighted or reversible (e.g., for different game)
138 R	CHANCE DEVICES
138 A	..Electric or magnetic
139	.Chance selection
140	..Fish ponds
141 R	.Rotating pointer
141 A	...Electric or magnetic
142 R	.Rotating disk
142 A	...Indicator: projected
142 B	...Indicator: electrical
142 C	...Indicator: optional
142 D	...Indicator: free element
142 E	...Ball indicator-pocketed disc
142 F	...Ball indicator-notched disc
142 G	...Ball indicator-stationary pocket
142 H	...Plural disc
142 HA	....Concentric
142 J	...Indexing
142 JA	....Pin and reed
142 JB	....Magnetic
142 JC	....Leaf spring and cog
142 JD	....Pivoted pawl
142 K	...Phonograph mounted
143 R	..Edge indication
143 A	....Rolling
143 B	....Movable web
143 C	....Indexing: magnetic
143 D	....Indexing: leaf spring
143 E	....Indexing: loose weight
144 R	.Lot mixers and dispensers
144 A	...Lot dispenser: mixing and dispensing
144 B	...Lot dispenser: mixing and showing
145 R	..Dice agitators
145 A	....Cup
145 B	....Chute
145 C	....Closed container
145 CA	.....Internal agitator
145 D	....Reciprocating floor
145 E	....Rotating table
146	.Dice
147	.Tops
292	CARD OR TILE GAMES, CARDS OR TILES THEREFOR
293	.Card or tile structure
294	..Playing surface having nonrectangular perimeter
295	..Material
296	.With functional back indicia
297	.Property or commodity transaction representation
298	.Sports or outdoor recreational activities
299	.Word, sentence, or equation forming
300	.With quotation thereon
301	.With musical indicia
302	.With educational data
303	.Suits
304	..With supplementary indicia
305	..Indexing
306	..With auxiliary or accessory card or tile
307	.Rearranged basic indicia
308	.With representations of persons or objects and names associated therewith
148 R	GAMES ACCESSORIES
148 A	..With card holders
148 B	..All video game accessories
309	.Game supporting tables or surfaces
149 R	.Card shufflers and dealers
149 P	...Devices for dealing predetermined hands
150	.Hand holders
151	..Duplicate games
153 R	PUZZLES
153 P	..Pyramid building
153 S	..Shifting movement
153 J	..Jumping movement
154	.Balancing ovoids
155	.Folding and relatively movable strips and disks
156	.Take-aparts and put-togethers
157 R	..Geometrical figures, pictures, and maps
157 A	....Transparent overlay
158	..Bent wire
159	..Flexible cord or strip
160	..Mortised blocks
161	FORTUNE-TELLING DEVICES

****

DIGESTS

****

DIG 1	Carbonate
DIG 2	Styrene
DIG 3	Epoxy
DIG 4	Ethylene
DIG 5	Vinyl
DIG 6	Nylon
DIG 7	Glass fiber
DIG 8	Urethane
DIG 9	Ester
DIG 10	Butadiene
DIG 11	Acetal
DIG 12	Propylene
DIG 13	Artificial grass
DIG 14	Transparent
DIG 15	Cork
DIG 16	Acrylic
DIG 17	Head mounted
DIG 18	Shoe mounted
DIG 19	Waist mounted
DIG 20	Weighted balls
DIG 21	Reel
DIG 22	Ionomer
DIG 23	High modulus filaments
DIG 24	Luminescent, phosphorescent
DIG 25	Suction cups involved
DIG 26	Point counters and score indicators
DIG 27	Blind and color blind
DIG 28	Cathode ray tube
DIG 29	Silicone
DIG 30	Hooked pile fabric fastener, e.g., VELCRO
DIG 31	Undulated surface

D25-1

CLASS D25 BUILDING UNITS AND CONSTRUCTION ELEMENTS

JUNE 1990

Excluded from this class are:
Burial vault; see D99-1+.
Cloth sheet material; see D5.
Concrete wall tie; see D8-384.
Diving board; see D21-236.
Elevator; see D34-33.
Escalator; see D34-28.
Fence or railing hardware, e.g., mounting bracket, clip, etc.; see D8.
Fireplace mantel; see D23-404.
Form for flower arrangement; see D11-143+.
Gate hinge; see D8-323+.
Gate latch; see D8-331+.
Grille for audio or video equipment; see D14-219.
Lamp combined with post; see D26-68.
Lens or glass for lighting fixture; see D26-120+.
Linoleum rug; see D6-582+.
Linoleum sheet; see D5.
Mobile home or trailer; see D12-101+.
Mobile stair for boarding aircraft, etc.; D34-30.
Pipe, tubing or hose; see D23-266.
Plant stake; see D8-1.
Portable swimmimg pool; see D21-252.
Roof gutter; see D23-267.
Service station canopy combined with pump; see D15-9.1+.
Step plate for vehicle; see D12-203.
Structure which encloses telephone only; see D6-421 and 553+.
Tent; see D21-253.
Tiles assembled into countertop; see D23-308.
Tiles assembled into floor and wall covering; D6-582+.
Track for sliding door; see D8-377.
Vehicle support; see D34-31.
Wallpaper; see D5.

1 STRUCTURE (1)
2 .Swimming pool
3 .Combined with diverse article, e.g., airport complex, etc.
4 .Clustered identical units
5 .Four or more levels
6 ..Tower
7 .Simulative
8 ..Animate
9 ...Humanoid
10 ..Edible article or container therefor
11 ..Vehicle
12 .Arena or stadium
13 .Geodesic dome (2)
14 .Grain storage
15 .Greenhouse
16 .Single occupant type, i.e., telephone booth, privy, ticket booth, etc. (3)
17 .Having intersecting angular or curved roofs (4)
18 .Having curved or vaulted roof
19 ..Dome
20 ..Sinuous
21 ..Concave, catenary or flexibly suspended
22 .Having angular roof
23 ..Trapezoidal or triangular surface
24 ...Gambrel type
25 ...Mansard type
26 ...Pyramidal
27 ...Zig-zag type, i.e., sawtooth
28 ...Inverted
29 ...A-frame type
30 ...Isolated slanted surface
31 .Circular or oval in plan
32 .Polygonal in plan
33 .Fully enclosed
34 ..Having carport or garage
35 PREFABRICATED UNIT (5)
36 .Underground service vault (6)
37 .Elevator cab interior (7)
38 .Railing unit or fence (8)
39 ..Combined, e.g., fence and gate
40 ..Simulative
41 ..Swimming pool grab rail
42 ..Post or board extending above rail, e.g., picket type
43 ..Post and rail type
44 ...Rail centered on post
45 ..Wire type (9)
46 ...Barbed
47 .Closure (10)
48 ..For passage, e.g., door, etc.
49 ...Flexible or folding
50 ...Gate
51 ....Turnstile
52 ..Window
53 ..Screen or guard (11)
54 ...Window well cover
55 .Finishing unit (12)
56 ..Roof or canopy (13)
57 ...Awning type
58 ..Partition, wall or ceiling (25)
59 ...Exterior, e.g., storefront
60 ..Surround for door window, panel, etc. (14)
61 .Framing unit, e.g., building frame (15)
62 STAIR, LADDER, SCAFFOLD OR SIMILAR SUPPORT (16)
63 .Stepped (17)
64 ..Ladder
65 ...Step stool type
66 .Scaffold
67 .Sawhorse
68 .Element or attachment (18)
69 ..Stair tread, step or ladder rung
100 TRELLIS OR TREILLAGE UNIT
101 .Simulative
102 ARCHITECTURAL STOCK MATERIAL (19)
103 .Transparent or translucent
104 ..Simulative
105 ...Plant life
106 ..Three or more pattern repeats about axis on side (20)
107 ...Hollow type
108 ..Hollow type
109 ..Symmetrical on two axes on one side (21)
110 ..Unidirectional repeat pattern
111 ..Random arrangement of pattern elements within repeat unit (22)
112 .Splash block
113 .Construction block or brick type (23)
114 ..Perforated or openwork (24)
115 ...Symmetrical
116 ....Arcuate detail
117 .....Cicular or oval
118 ....Rectangular or square detail
119 .Extruded shape (34)
120 ..Three or more repeats, or uniform configuration, about axis
121 ..Bilaterally symmetrical
122 ...With hollow core
123 ...Parallel repeating ribs or grooves
124 ..With hollow core
125 ..Parallel repeating ribs or grooves
126 .Post or beam
127 ..With cross arm or cantilevered support
128 ..Three or more repats, or uniform configuration, about axis
129 ...Vertically fluted
130 ..Simulative

D25-2

CLASS D25 BUILDING UNITS AND CONSTRUCTION ELEMENTS

JUNE 1990

ARCHITECTURAL STOCK MATERIAL (19)
.Post or beam
131 ..With notch or tang for receiving attachment (e.g., fencing wire, etc.)
132 ..Openwork type or with opening
133 ..Element or attachment
134 ...Cross arm or cantilevered support
135 ...Capital or cap
136 .Elongated molding type (e.g., crown, dentil, etc.) (26)
137 ..Simulative
138 .Panel, tile or applique or floor, wall, roof, etc. (27)
139 ..Roofing, or wall shingle
140 ...Tile type
141 ...Stamped, crimped or bent material having uniform thickness
142 ...Perforated or openwork
143 ...Three or more repeats, or uniform configuration, about axis
144 ...Symmetrical on two axes
145 ..Relief carving or medallion type (28)
146 ...Humanoid simulation
147 ...Three or more repeats, or uniform configuration, about axis (e.g., medallion, etc.)
148 ...Symmetrical on two axes
149 ..Simulative, or natural material pattern
150 ...Wood grain
151 ...Brick or stone (29)
152 ..Perforated or openwork (30)
153 ...Symmetrica pattern or pattern unit
154 ....Arcuate
155 .....Circular or oval
156 ....Rectangular or square
157 ..Symmetrical pattern or pattern unit (31)
158 ...Three or more repeats, or uniform configuration, about axis
159 ....Consisting entirely of material having uniform thickness
160 ...On two axes
161 ....Consisting entirely of aterial having uniform thickness
162 ...Consisting entirely of material having uniform thickness
163 ..Textured (e.g., pitted, cratered, drilled, etc)
164 .Elongated (32)
199 MISCELLANEOUS (33)

SEARCH NOTES FOR CLASS D25.

(1) Includes bridge.
For roof with no structure other than support, see subclasses 56+.
(2) For arcuate dome, see subclass 19.
(3) Includes structures which enclose a person.
(4) Search subclasses 56+ for roof, per se.
(5) Includes only plural components which are preassembled as a unit, e.g., steeple, cupola, etc.
For individual components, see subclasses 102+.
For trellis, stairs, louver unit, see, respectively, subclasses 100, 62+ and 152+.
For girder, beam, etc., see subclasses 126+.
(6) Includes element.
(7) For elevator, see D34-33.
(8) For handrail or post, see respectively, subclasses 119 and 126+.
(9) Includes element.
(10) Includes shutter; includes a combination with surround.
For surround per se, see subclass 60.
(11) Includes only screens or guards which are designed solely for window or door.
For grille adaptable to other architectural uses, see subclasses 152+.
For perforated sheet material, see D5-1+.
(12) Includes rigid drapery cornice.
(13) If structure has wall, see subclasses 1+.
(14) Includes arch structure.
(15) Includes cantilever support.
(16) Includes ramp; includes bleachers.
(17) For mobile stairs, see D34-30.
(18) For component which is adaptable to other article, such as handrail, post, etc., see subclasses 102+.
(19) Classified by full line disclosure
(20) Planar surface not considered.
(21) Planar surface not considered.
(22) If repeat protocol is apparent from the disclosure, patent is classified under earlier subclasses. If, however, the disclosured pattern element does not indicate how it can repeat, patent is classified here.
(23) For edging, see subclass 164.
(24) Opening must extend through the article.
(25) Includes stage set.
(26) If molding has extruded shape, see subclasses 119+.
For corner or end piece, see subclass 102.
(27) For generic elongated stock material having indefinite length, see subclass 164.
(28) For embossed panelwork or for grill, see subclasses 138, 152 or 157.
(29) For brick per se, see subclasses 1
(30) Opening must extend through the article. Otherwise patent is classified in subclasses which follow.
(31) Exclusive of random acoustic texture within symmetrical pattern unit, or of tongue or groove edge detail.
(32) Includes railing, edging, etc. However, if railing, etc. has an extruded shape, patent is classified in subclasses 119+.
For corner or end piece, see subclass 102.
(33) Includes construction form.
Includes roof pipe flashing.
(34) To be classified in this subclass, article must have uninterrupted extruded shape. If openings, protrusions, rivets, vertical striations, etc., are disclosed with an elongated article, patent is classified in subclasses 126, 136 or 164.
For articles with specific end configuration such as fence picket, see subclasses 126+.

☆ U.S. GOVERNMENT PRINTING OFFICE 1990 — 261-920 20629

CLASS PLT PLANTS

PLT-1

DECEMBER 1986

1 ROSES
2 .Climbers
3 ..White or yellow
4 ..Orange or salmon
5 ..Pink
6 ..Red
7 .Miniatures
8 ..White or yellow
9 ..Pink
10 ..Red
11 .Hybrid teas or grandifloras
12 ..Red bicolor
13 ..Striped colors
14 ..White
15 ..Yellow
16 ..Orange
17 ..Salmon
18 ..Light to medium pink
19 ..Dark pink
20 ..Light to medium red
21 ..Dark red
22 .Floribundas or polyanthas
23 ..White
24 ..Yellow
25 ..Salmon
26 ..Light to medium pink
27 ..Dark pink
28 ..Light to medium red
29 ..Dark red
30 NUTS (INCLUDING ORNAMENTAL VARIETIES)
31 .Pecans
32 .Walnuts
33 FRUITS (EDIBLE AND FLESHY, BUT ALSO INCLUDING ORNAMENTAL VARIETIES)
34 .Apples (including crabapples)
35 ..Sports of "Delicious"
36 .Pears
37 .Cherries
38 .Plums or prunes
39 .Apricots
40 .Nectarines
41 ..Yellow-fleshed
42 .Peaches
43 ..Yellow-fleshed
44 .Avocados
45 .Citrus
46 .Brambles (i.e., the genus Rubus)
47 .Grapes
48 .Strawberries
49 ..Everbearing
50 CONIFERS
51 BROADLEAF TREES
52 .Honey locusts
53 .Poplars
54 SHRUBS OR VINES
55 .Azaleas or rhododendrons
56 ..Light to medium pink
57 ..Dark pink to red
58 .Barberries
59 .Buddleias
60 .Camellias
61 ..Light to medium pink
62 ..Dark pink to red
63 .Euonymus
64 .Flowering quinces
65 .Hollies
66 .Lilacs
67 .English ivies (i. e., Hedera helix varieties)
68 HERBACEOUS FLOWERING PLANTS
69 .African violets
70 .Carnations or pinks
71 ..Light to medium pink
72 ..Dark pink
73 ..Red
74 .Chrysanthemums
75 ..Cushion
76 ..Decorative (i.e., double-flowered)
77 ...White
78 ...Yellow
79 ...Orange
80 ...Light to medium pink
81 ...Dark pink
82 ...Red
83 .Freesias
84 .Fuchsias
85 .Gladioli
86 .Poinsettias
87 .Verbenas
88 HERBACEOUS ORNAMENTAL FOLIAGE PLANTS (INCLUDING LAWN GRASSES)
89 MISCELLANEOUS (E.G., MINTS, HOPS, MUSHROOMS, SUGAR CANE OR TOBACCO)

PLT-1

CLASS PLT PLANTS

DECEMBER 1986

1 ROSES
2 .Climbers
3 ..White or yellow
4 ..Orange or salmon
5 ..Pink
6 ..Red
7 .Miniatures
8 ..White or yellow
9 ..Pink
10 ..Red
11 .Hybrid teas or grandifloras
12 ..Red bicolor
13 ..Striped colors
14 ..White
15 ..Yellow
16 ..Orange
17 ..Salmon
18 ..Light to medium pink
19 ..Dark pink
20 ..Light to medium red
21 ..Dark red
22 .Floribundas or polyanthas
23 ..White
24 ..Yellow
25 ..Salmon
26 ..Light to medium pink
27 ..Dark pink
28 ..Light to medium red
29 ..Dark red
30 NUTS (INCLUDING ORNAMENTAL VARIETIES)
31 .Pecans
32 .Walnuts
33 FRUITS (EDIBLE AND FLESHY, BUT ALSO INCLUDING ORNAMENTAL VARIETIES)
34 .Apples (including crabapples)
35 ..Sports of "Delicious"
36 .Pears
37 .Cherries
38 .Plums or prunes
39 .Apricots
40 .Nectarines
41 ..Yellow-fleshed
42 .Peaches
43 ..Yellow-fleshed
44 .Avocados
45 .Citrus
46 .Brambles (i.e., the genus Rubus)
47 .Grapes
48 .Strawberries
49 ..Everbearing
50 CONIFERS
51 BROADLEAF TREES
52 .Honey locusts
53 .Poplars
54 SHRUBS OR VINES
55 .Azaleas or rhododendrons
56 ..Light to medium pink
57 ..Dark pink to red
58 .Barberries
59 .Buddleias
60 .Camellias
61 ..Light to medium pink
62 ..Dark pink to red
63 .Euonymus
64 .Flowering quinces
65 .Hollies
66 .Lilacs
67 .English ivies (i. e., Hedera helix varieties)
68 HERBACEOUS FLOWERING PLANTS
69 .African violets
70 .Carnations or pinks
71 ..Light to medium pink
72 ..Dark pink
73 ..Red
74 .Chrysanthemums
75 ..Cushion
76 ..Decorative (i.e., double-flowered)
77 ...White
78 ...Yellow
79 ...Orange
80 ...Light to medium pink
81 ...Dark pink
82 ...Red
83 .Freesias
84 .Fuchsias
85 .Gladioli
86 .Poinsettias
87 .Verbenas
88 HERBACEOUS ORNAMENTAL FOLIAGE PLANTS (INCLUDING LAWN GRASSES)
89 MISCELLANEOUS (E.G., MINTS, HOPS, MUSHROOMS, SUGAR CANE OR TOBACCO)

Appendix 20
Search Screens

F1:Help F2:Full List F3:Format (Sort) F5:Output F6: Jump F7:Done

Patent Number	Patent Title — Patent Title Listing: 1 of 4
4666162	CONSTRUCTION GAME
4387897	GAME APPARATUS
4325552	MANIPULATIVE TOY
4043561	PUZZLE AND METHOD

CASSIS CD-ROM search showing patent numbers housed in class 273 subclass 153p

F1:Help F2:Full/List F3:Format (Sort) F5:Output F6:Jump F7:Done

Patent Display: 1 of 4

Patent Number	4666162
Issue Year	987
Assignee Code	0
State / Country	CAX
Classification	273/249 273/241 273/153P 273/256 273/276
Title	CONSTRUCTION GAME APPARATUS
Abstract	A game apparatus includes a horizontal playing surface divided into individual playing areas vertical intersecting walls. Play involves the assembly of geometrical building blocks within each playing area to progressively construct an identical segment on a building structure that is symmetrical across the walls and that includes the walls themselves. The building blocks must be arranged or predetermined layers of diminishing width and length to form a pyramidal structure. The upper wall surfaces are configured as stairways, causing the completed building structure to have the appearance of a Mayan temple.

Abstract demonstrating patent resident in an alpha subclass

F1:Help F2:Browse F3:Display F4:Connection F5: Storage F6:Setup F7:Quit

SEARCH

U.S. Department of Commerce
Patent and Trademark Office
PATENT BIBLIOGRAPHIC FILE - 1969 TO DATE

Patent Number:
Issue Year: 4
Assignee Code:
State or Country:
Status:
Classification: 273/153p
Title or Abstract:

Use arrow keys to highlight search field.
Touch ENTER to start search, CTRL-BREAK to abort search.

Connection: Total: 4

CASSIS CD-ROM search on class 273 subclass 153p demonstrating alpha classes

F1:Help F2:Browse F3:Display F4:Connection F5: Storage F6:Setup F7:Quit

SEARCH

U.S. Department of Commerce
Patent and Trademark Office

PATENT BIBLIOGRAPHIC FILE - 1969 TO DATE

Patent Number:
Issue Year:
Assignee Code:
State or Country:
Status:
Classification: 273/153p
Title or Abstract:

4

Use arrow keys to highlight search field.

Touch ENTER to start search, CTRL-BREAK to abort search.

Connection: Total: 4

CASSIS CD-ROM search on class 273 subclass 153p demonstrating alpha classes

Appendix 21
Sample Patents from the *Official Gazette*

4,496,657
MICROPLATE WASHER
Ward J. Coppersmith, San Diego, and George B. LaMotte, III, Larkspur, both of Calif., assignors to Scripps Clinic and Research Foundation, Calif.
Filed Oct. 14, 1982, Ser. No. 434,308
Int. Cl.^{3} C12M 1/00, 1/32; B01L 3/00
U.S. Cl. 435—287 10 Claims

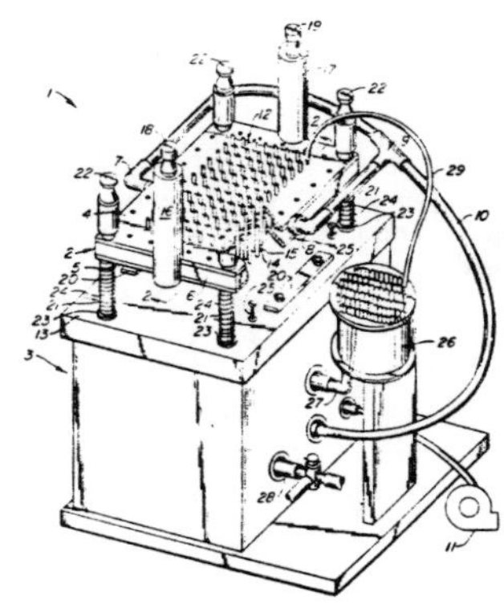

1. An apparatus for the alternate filling and evacuation of all wells simultaneously of a multiwell microplate, said apparatus comprising:

a vacuum chamber defined by two parallel plates designated an upper plate and a lower plate, respectively, said plates separated by peripheral spacing means, said lower plate having affixed therein a plurality of tubes spaced apart from each other, each said tube passing through said lower plate and perpendicular thereto, said tubes forming a rectangular array such that, when said chamber is superimposed over said microplate, a pair of said tubes is aligned with each well of said microplate, the portions of each tube of said pair which extend below said lower plate being of unequal length, the tube with the shorter portion below the lower plate being designated a filling tube and the tube with the longer portion below the lower plate being designated an evacuation tube, said filling tube extending upward through said chamber and said upper plate, and said evacuation tube terminating within said chamber above the lower plate,

a support for said microplate, said support comprising a substantially flat horizontal surface containing guide means adapted to secure said microplate in a predetermined position on said support surface,

means for maintaining said vacuum chamber in vertical alignment with said support with said upper and lower plates of said vacuum chamber and said support surface in parallel relation, and for narrowing and widening the

distance between said vacuum chamber and said support surface while substantially maintaining said parallel relation,
stop means for setting a minimum spacing distance between said vacuum chamber and said support surface such that when said distance is at said minimum, the exposed end of each said evacuation tube is sufficiently close to the bottom of the corresponding well in said microplate to permit evacuation of substantially all liquid from said well,
means for applying a vacuum to said vacuum chamber, and
means for supplying pressurized fluid to each said filling tube at the upper end thereof.

277,377
TELEPHONE SET
Masaharu Tarao, and Katsuhito Watanabe, both of Tokyo, Japan, assignors to Oki Electric Industry Co., Ltd., Tokyo, Japan
Filed Jun. 17, 1982, Ser. No. 378,488
Claims priority, application Japan, Jan. 12, 1982, 57-00505
Term of patent 14 years
U.S. Cl. D14—53

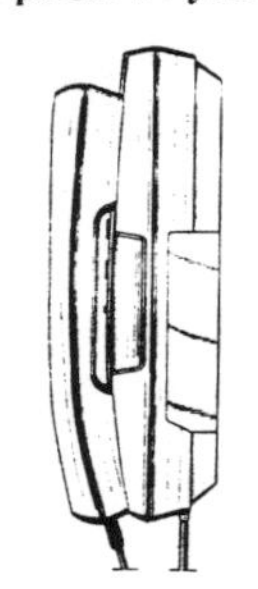

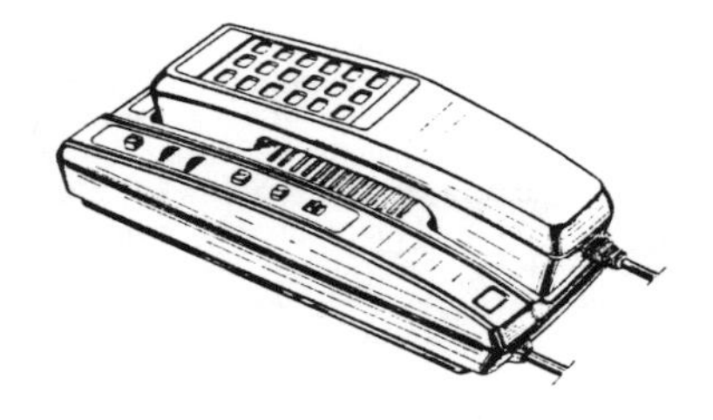

Illustrations for plant patents are usually in color and therefore it is not practicable to reproduce the drawing.

5,396
CLIMBER ROSE PLANT—METZALITAF VARIETY
Marie L. Meilland, Antibes, France, assignor to The Conard-Pyle Company, West Grove, Pa.
Filed Jun. 28, 1983, Ser. No. 508,717
Int. Cl.[3] **A01H** *5/00*
U.S. Cl. Plt.—2 1 Claim

1. A new and distinct variety of Climber rose plant, substantially as illustrated and described, characterized particularly by (a) a vigorous well branched growth habit with long arching canes, (b) buds which open into extremely attractive relatively large blossoms of a richly colored yellow-rouge blend, and (c) attractive leathery and glossy foliage of deep green coloration.

Appendix 22
List of Patentees from *Index of Patents*

INDEX OF PATENTS

Part I, List of Patentees

ISSUED FROM

THE UNITED STATES
PATENT OFFICE

1973

U.S. GOVERNMENT PRINTING OFFICE
WASHINGTON : 1974

For sale by the Superintendent of Documents, U.S. Government Printing Office, Washington, D.C. 20402
Price $19.80 (Paper Cover) Stock Number 0304-00507

CONTENTS

1973

PATENTS ISSUED[1]

Patents	74,139	—No. 3,707,729 to No. 3,781,913 inclusive.
Reissues	314	—No. 27,540 to No. 27,855 inclusive.
Designs	4,033	—No. 225,695 to No. 229,728 inclusive.
Plant Patents	132	—No. 3,281 to No. 3,412 inclusive.
Defensive Publications	143	—No. T908,001 to No. T917,012.

[1] The number of patents issued during a year may be fewer than the number derived by subtracting the number of the first patent and the last patent due to withdrawals, etc.

Lundquist, Richard E.; and Kerr, William J., 3,745,430.
Kerr, William J., to Chicago Lock Co. Axial pin tumbler lock assembly and combination reset key therefor. 3,756,049, 9-4-73, Cl. 70-363.000.
Kerr-McGee Chemical Corporation: *See*—
Bradford, James Lewis, 3,767,776.
Rhees, Raymond C., 3,781,412.
Schlaudroff, Leo M., 3,778,890.
Kerr-McGee Corporation: *See*—
Baldwin, Roger A.; and Cheng, Ming T., 3,714,267.
Baldwin, Roger A.; and Cheng, Ming T., 3,761,527.
Chappelow, Cecil C., Jr.; and Engel, James F., 3,742,062.
Eberline, Howard C., 3,773,109.
Lucid, Michael F.; and Leader, William M., 3,734,696.
Lucid, Michael F., 3,764,274.
Kerridge, Mark L.: *See*—
Putnam, Allen Lewis, 3,724,119.
Kerrigan, Charles M.: *See*—
Hai, Atta Mohammed; Kerrigan, Charles M.; and Leidy, Harold T., 3,719,499.
Leidy, Harold T.; Kerrigan, Charles M.; Tewey, Robert T.; and Bartenbach, Louis, 3,711,291.
Leidy, Harold T.; Kerrigan, Charles M.; and Byble, Duane C., 3,713,837.
Leidy, Harold T.; Kerrigan, Charles M.; Tewey, Robert T.; and Bartenbach, Louis, 3,719,498.
Kerrigan, James E.; and Klygis, Mindaugas J., to Illinois Tool Works, Inc. Container package. 3,734,278, 5-22-73, Cl. 206-65.00s.
Kersch Company: *See*—
Ford, James A.; and Ford, Alan A., 3,766,597.
Kerschbaum, Ewald: *See*—
Baden, Werner Fidi; and Kerschbaum, Ewald, 3,739,195.
Kerschner, James J., to Acme Highway Products Corporation. Sealing member. 3,718,403, 2-27-73, Cl. 404-49.000.
Kerschner, Paul M.: *See*—
Badin, Elmer J.; Kerschner, Paul M.; and Cresti, Aldo, 3,765,850.
Kerschner, Paul M.; and Marcellis, Alphonso W., to Cities Service Oil Company. Petroleum hydrocarbon compositions. 3,761,434, 9-25-73, Cl. 260-22.00r.
Kerschner, Paul M.; Badin, Elmer J.; and Cresti, Aldo, to Cities Service Oil Company. Nitrogen-containing carbohydrate derivatives and hydrocarbon fuel compositions containing same. 3,779,724, 12-18-73, Cl. 44-63.000.
Kerschner, Paul M., to Cities Service Company. Polybromocycloaliphatic ethers as flame retardant plasticizers. 3,779,978, 12-18-73, Cl. 260-33.20r.
Kersey, John P.: *See*—
Flournoy, Norman E.; and Kersey, John P., 3,769,711.
Kershaw, Bernard John, to Du Pont of Canada Limited. Purification of adiponitrile. 3,775,258, 11-27-73, Cl. 203-29.000.
Kershaw, Joseph E., to United States of America, Navy. Method and apparatus for coupling multiple power sources to single radiating antenna. 3,714,661, 1-30-73, Cl. 343-858.000.
Kershaw, Robert William; and Polgar, Livia, to Dulux Australia Ltd. Liquid compositions comprising gelled urethane polymer. 3,748,294, 7-24-73, Cl. 260-22.0tn.
Kershaw, Samuel L.: *See*—
Bianchetta, Donald L.; Guhl, Richard E.; Kershaw, Samuel L.; and Szentes, John F., 3,727,730.
Kershaw, Stanley S., Jr.: *See*—
Marek, James R.; and Kershaw, Stanley S., Jr., 3,710,212.
Kershaw, Sydney L., to Parker-Hannifin Corporation. Quick disconnect coupling. 3,758,137, 9-11-73, Cl. 285-70.000.
Kershner, Osborn A., to Lambert Brake Corporation, mesne. Self-adjusting disc brake assembly. 3,722,637, 3-27-73, Cl. 188-196.00p.
Kershner, Stephen W.; and Owen, Kenneth, to Delta Electronics, Inc. Balanced line switch system with U-shaped crossed conductive channels arranged back to back. 3,717,736, 2-20-73, Cl. 200-153.00s.
Kerst, Al F., to Monsanto Company. Methods of scale inhibition. 3,733,270, 5-15-73, Cl. 210-58.000.
Kerst, Al F.; and Douros, John D., Jr., to Gates Rubber Company, The. 4,5 Diaminouracil sulfate as algicide. 3,753,362, 8-21-73, Cl. 71-67.000.
Kerst, Al F.; and Douros, John D., Jr., to Gates Rubber Company, The. Diethyl cyanomethyl phosphonate as an antimicrobial agent. 3,764,676, 10-9-73, Cl. 424-210.000.
Kerst, Al F.; and Douros, John D., Jr., to Gates Rubber Company, The. Diethyl beta aminoethylphosphonate as an antimicrobial agent. 3,764,677, 10-9-73, Cl. 424-211.000.
Kerst, Al F.; Douros, John D., Jr.; and Brokl, Milan, to Gates Rubber Company, The. Controlling algae with 5-(5 barbiturilidene)-rhodanine. 3,765,864, 10-16-73, Cl. 71-67.000.
Kerst, Al Fred: *See*—
Douros, John D., Jr.; and Kerst, Al Fred, 3,725,557.
Douros, John D., Jr.; and Kerst, Al Fred, 3,728,453.
Douros, John D., Jr.; and Kerst, Al Fred, 3,728,454.
Douros, John D., Jr.; and Kerst, Al Fred, 3,728,461.
Douros, John D., Jr.; Broke, Milan; and Kerst, Al Fred, 3,773,952.
Kersten, Hilde; Heinrichs, Gunter; Meyer, Gerhard; and Laudien, Dieter. Process for preparing 1,3-disubstituted symmetrical thioureas. 3,708,496, 1-2-73, Cl. 260-309.700.
Kersten, Hilde; Meyer, Gerhard; and Neuhaus, Clemens, to Akzona Incorporated. Production of thiourea. 3,723,522, 3-27-73, Cl. 260-552.00r.
Kerstetter, Donald R.: *See*—
Bingeman, Wilbur H.; Grimone, Frank H.; and Kerstetter, Donald R., 3,767,958.
Buescher, William E.; and Kerstetter, Donald R., 3,730,706.
Decker, John J.; and Kerstetter, Donald R., 3,737,342.
Kerstetter, Donald R.; and Losey, Harold D., to Sylvania Electric Products, Inc. Vacuum tube readout device having ruggedized internal electrode structure and method of making same. 3,716,899, 2-20-73, Cl. 29-25.140.
Kerstetter, Donald R., to GTE Sylvania Incorporated. Electron discharge device grid having enhanced thermal conductivity and reduced secondary emission characteristics. 3,772,561, 11-13-73, Cl. 313-350.000.
Kerstetter, Harold Alfred: *See*—
Crumley, J. A.; Hildebrand, James Ross; Kerstetter, Harold Alfred; and Shaak, Ray Ned, 3,712,735.
Kersting, Arno: *See*—
Keller, Wolfgang; Kersting, Arno; and Reuschel, Konrad, 3,781,152.
Reuschel, Konrad; Kersting, Arno; and Keller, Wolfgang, 3,751,539.
Kersting, Raymond J.: *See*—
Hardwick, David R.; and Kersting, Raymond J., 3,744,848.
Kersting, Raymond J., to Wagner Electric Corporation. Control valve. 3,712,686, 1-23-73, Cl. 303-71.000.
Kersting, Raymond J., to Wagner Electric Corporation. Control valve. 3,778,119, 12-11-73, Cl. 303-68.000.
Kersting, Raymond J., to Wagner Electric Corporation. Control valve. 3,781,065, 12-25-73, Cl. 303-71.000.
Kerswill, Edson G.: *See*—
Urbanek, Karel; and Kerswill, Edson G., 3,741,886.
Kertell, Frank W., to Walker, Brooks. Quad jet. 3,758,082, 9-11-73, Cl. 261-23.00a.
Kerttula, Into Isak; and Jaatinen, Per Arno. Device in belt conveyors. 3,729,089, 4-24-73, Cl. 198-203.000.
Kerttula, Into Isak; and Jaatinen, Per Arno. Continuous action sheet press. 3,779,686, 12-18-73, Cl. 425-371.000.
Kertzman, Jack: *See*—
Skarstrom, Charles W.; and Kertzman, Jack, 3,735,558.
Kertzman, Norman. Compartmented portable case. 3,739,886, 6-19-73, Cl. 190-49.000.
Kervalishvili, Nikolai Arsenovich: *See*—
Barkhudarov, Eduard Mikhailovich; Kervalishvili, Nikolai Arsenovich; and Kortkhondzhia, Vladimir Parmenovich, 3,728,246.
Kervizic, Jacques; Masic, Rene; and Warnecke, Robert Jean, to Thomson-CSF. Turret device for positioning crucibles in ion sources. 3,770,870, 11-6-73, Cl. 13-31.000.
Kerwin, Robert Eugene: *See*—
Maydan, Dan; Cohen, Melvin Irwin; and Kerwin, Robert Eugene, 3,720,784.
Kerwood, Joseph Edward: *See*—
Coran, Aubert Yaucher; and Kerwood, Joseph Edward, 3,752,824.
Coran, Aubert Yaucher; and Kerwood, Joseph Edward, 3,775,428.
Kerwood, Joseph Edward, to Monsanto Company. N-azolylsulfenamides. 3,770,758, 11-6-73, Cl. 260-305.000.
Kerzman, Jack A. Wad assembly for shotgun shell. 3,707,915, 1-2-73, Cl. 102-42.00c.
Keserin, Ivan: *See*—
Koy, Hermann; and Keserin, Ivan, 3,745,620.
Keserin, Ivan, to Metallgesellschaft Aktiengesellschaft. Corona-discharge electrode mounting. 3,775,827, 12-4-73, Cl. 29-275.000.
Kesinger, Donald A.; and Inscho, Leland S., Jr., to Gates Rubber Company, The. Transplant handling means. 3,722,137, 3-27-73, Cl. 47-34.130.
Keskilohko, Altti Kalervo: *See*—
Bergius, Mikko Hannu Tapani; Saarenketo, Tapio Keikki; and Keskilohko, Altti Kalervo, 3,749,034.
Kesler, George H.: *See*—
Carpenter, James F.; McCrary, Leon E.; Steube, Kenneth E.; Klein, Albert A., Jr.; and Kesler, George H., 3,750,623.
Kesler, Richard B., to Institute of Paper Chemistry, The. Method and apparatus for analysis of fluid suspensions. 3,718,030, 2-27-73, Cl. 73-61.00r.
Kesler, Sidney B., Jr., to Reynolds Metals Company. Apparatus for removing treating liquids from treated metal strip products. 3,778,862, 12-18-73, Cl. 15-306.400.
Kesling, Keith K., to General Motors Corporation. Ice flaking machine for domestic refrigerators. 3,727,425, 4-17-73, Cl. 62-346.000.

Appendix 23
Patent for Styrofoam Cups

2,828,509
PLASTIC MOLDING MACHINES
Robert E. Smucker and James M. Harrison, Fort Worth, Tex., assignors to Crown Machine and Tool Company, Fort Worth, Tex., a corporation of Texas
Application November 3, 1954, Serial No. 466,482
5 Claims. (Cl. 18—30)

1. In a molding machine for molding relatively thin walled cup-shaped plastic articles and the like, a frame, a cavity platen on the frame, a cavity insert in the cavity platen having a cavity, a core platen on the frame with a core element insertable into the cavity to define a relatively thin walled cup-shaped molding cavity, an injection nozzle on the frame in communication with the molding cavity for supplying liquid plastic under pressure to the molding cavity, and a coolant course for the molding cavity for circulating a coolant including a plurality of separate closed channels between the cavity platen and cavity insert around the molding cavity interconnected by staggered axial passages, the channels being relatively closely spaced to the molding cavity so that heat in the plastic injected into the molding cavity may be rapidly withdrawn by the coolant flowing through the channels,

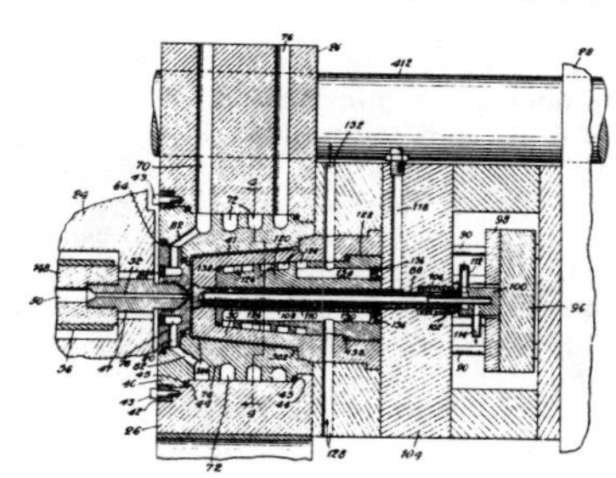

an inlet communicating with one such closed channel at one end of the plurality, and an outlet communicating with another such closed channel at the other end for circulation of the coolant.

Appendix 24
Patent for Ledger Paper

1,682,514. MEANS AND METHOD FOR PRODUCING VERTICALLY-RULED PRINTING FORMS. WILLIAM C. HOLLISTER and LEO M. CHAPMAN, Chicago, Ill., assignors to Chicago Lino-Tabler Company, Chicago, Ill., a Corporation of Illinois. Filed Mar. 7, 1925. Serial No. 13,734. 29 Claims. (Cl. 164—91.)

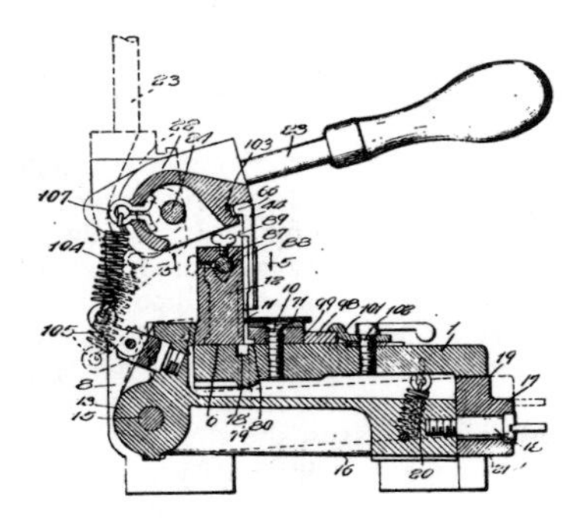

3. In a machine of the kind described, a series of punch units spaced apart and clamped in upright position, each unit comprising a rigid holder and a punch movable up and down in the holder, the punch provided with a rearwardly extending projection at its upper end, a punch operator on the machine movable up and down and having a part adapted to be pressed down upon the upper end of the punch to force the punch down, and provided with a ledge below its top for engagement beneath the rearwardly extending projection on the punch for lifting same, said operator being bodily movable forward and back for engaging and disengaging the punch, said operator arranged to be swung up away from the punch when disengaged from same to permit the convenient adjustment of the punch unit in the machine.

Appendix 25
Sample Entries from *Index to the U.S. Patent Classification*

INDEX TO THE U.S. PATENT CLASSIFICATION

DECEMBER 1987

Provided By The
PATENT DEPOSITORY
LIBRARY PROGRAM

U.S. DEPARTMENT OF COMMERCE
C. William Verity, Secretary

U.S. Patent and Trademark Office
Donald J. Quigg
Assistant Secretary and
Commissioner of Patents and Trademarks

Alice B. Aderholt, Editor
Office of Documentation

Appendix 26
Samples of Schedules from *Manual of Classification*

365-1

CLASS 365 STATIC INFORMATION STORAGE AND RETRIEVAL

DECEMBER 1986

1	MAGNETIC BUBBLES
2	.Disposition of elements
3	..Lattice
4	.Decoder
5	.Logic
6	.Rotating field circuits
7	.Detectors
8	..Magnetoresistive
9	..Hall effect
10	..Optical
11	.Generators
12	..By splitting
13	.Plural interacting paths
14	..Closed loop
15	...Major-minor
16	..With switch at interacting point
17	...Idler switch
18	..Boundary
19	.Conductor propagation
20	..Including A.C. signal
21	..Three phase signals
22	.One's and zero's
23	.Plural direction propagation
24	..Nonsequential
25	.Velocity
26	..Turns
27	.Bias
28	..Variable
29	.Strip domain
30	.In-plane field (nonrotating)
31	.Different size bubbles
32	.Multiple magnetic layer
33	.Magnetic storage material
34	..Amorphous
35	.Guide structure
36	..Ion implantation
37	..Slots or rails
38	..Zigzag
39	..Overlays
40	...On opposite sides of storage medium
41	...Dots
42	...Wedges
43	...Chevrons
44	...Rectangular bars
45	ANALOG STORAGE SYSTEMS
46	.Resistive
47	.Thermoplastic
48	.Magnetic
49	ASSOCIATIVE MEMORIES
50	.Magnetic
51	FORMAT OR DISPOSITION OF ELEMENTS
52	HARDWARE FOR STORAGE ELEMENTS
53	.Shields
54	.Ground plane
55	.Magnetic
56	..Spacers
57	..Keeper
58	..Slot
59	..Embedded conductor
60	..Air gap
61	..Hairpin conductor
62	..Permanent magnet
63	INTERCONNECTION ARRANGEMENTS
64	.Optical
65	.Ferroelectric
66	.Magnetic
67	..Plural diagonal
68	..Tree
69	..Crossover
70	..Woven
71	.Negative resistance
72	.Transistors or diodes
73	RECIRCULATION STORES
74	.Magnetic
75	.Stepwise
76	.Delay lines
77	.Plural paths
78	PLURAL SHIFT REGISTER MEMORY DEVICES.
80	MAGNETIC SHIFT REGISTERS
81	.Bidirectional
82	.Two cells per bit
83	.SiPo/PiSo
84	.Core in transfer loop
85	.Continuous
86	..Plated wire
87	.Thin film
88	..Domain tip
89	.Logic
90	.Multiaperture cell
91	..Ladder
92	..With other type core
93	.Including delay means
94	READ ONLY SYSTEMS (I.E.. SEMIPERMANENT)
95	.With override (i.e., latent images)
96	.Fusible
97	.Magnetic
98	..Random core
99	..Random wiring
100	.Resistive
101	.Inductive
102	.Capacitative
103	.Semiconductive
104	..Transistors
105	..Diodes
106	RADIANT ENERGY
107	.Chemical fluids
108	.Liquid crystal
109	.Photoconductive and ferroelectric
110	.Electroluminescent and photoconductive
111	.Electroluminescent
112	.Photoconductive
113	.Amorphous
114	.Semiconductive
115	..Diodes
116	.Plasma
117	.Ferroelectric
118	.Electron beam
119	.Color centers
120	INFORMATION MASKING
121	.Polarization
122	..Magneto-optical
123	.Bragg cells
124	.Diffraction
125	..Holograms
126	.Thermoplastic
127	.Transparency
128	.Electron beams
129	SYSTEMS USING PARTICULAR ELEMENT
130	.Three-dimensional magnetic array
131	.Two magnetic cells per bit
132	.Different size cores
133	.Cells of diverse coercivity
134	.Continuous cells
135	..Elongated or bar-shaped cell
136	...Twisters
137	...Tubular
138	...Chain
139	...Plated wire
140	.Multiaperture cell
141	..Aperture plate
142	..Aperture with transverse axis
143	...Biax
144	..Same size apertures
145	.Ferroelectric
146	.Electrets
147	.Persistent internal polarization (PIP)
148	.Resistive
149	.Capacitors
150	..Inherent
151	.Molecular or atomic
152	..Nuclear induction or spin echo
153	.Electrochemical
154	.Flip-flop (electrical)
155	..Plural emitter or collector
156	..Complementary
157	.Magnetostrictive or piezoelectric

365-2

CLASS 365 STATIC INFORMATION STORAGE AND RETRIEVAL

DECEMBER 1986

	SYSTEMS USING PARTICULAR ELEMENT
158	.Magnetoresistive
159	.Negative resistance
160	.Superconductive
161	..Thin film
162	..Josephson
163	.Amorphous (electrical)
164	.Electrical contacts
165	..Coherer
166	..Relay
167	.Simulating biological cells
168	.Ternary
169	.Gunn effect
170	.Hall effect
171	.Magnetic thin film
172	..Isotropic
173	..Multiple magnetic storage lavers
174	.Semiconductive
175	..Diodes
176	..Silicon on sapphire (SOS)
177	..Bioplar and FET
178	..Ion implantation
179	..Plural emitter or collector
180	..Four layer devices
181	..Complementary conductivity
182	..Insulated gate devices
183	...Charge coupled
184	...Variable threshold
185	...Floating gate
186	..Single device per bit
187	..Three devices per bit
188	..Four or more devices per bit
189	READ/WRITE CIRCUITS
190	.For complementary information
191	.Signals
192	..Radio frequency
193	..Strobe
194	..Delay
195	..Inhibit
196	...Sense/inhibit
197	..Microwave
198	..Transmission
199	..Coincident A.C. signal with pulse
200	.Bad bit
201	.Testing
202	.Complementing/balancing
203	.Precharge
204	.Accelerating charge
205	.Flip-flop used for sensing
206	.Noise suppression
207	..Differential sensing
208	...Semiconductors
209	...Magnetic
210	....Reference or dummy element
211	..Temperature compensation
212	...Semiconductor
213	...Magnetic
214	..Particular wiring
215	.Optical
216	..Holographic
217	.Electron beam
218	.Erase
219	.SiPo/PiSo
220	.Parallel read/write
221	.Serial read/write
222	.Data refresh
223	.Bridge
224	.Eddy current
225	.Minor loop
226	POWERING
227	.Conservation of power
228	.Data preservation
229	..Standby power
230	ADDRESS
231	.Using selective matrix
232	..Magnetic
233	.Sync/clocking
234	.Optical
235	..Page memories
236	.Counting
237	.Electron beam
238	.Cartesian memories
239	.Sequential
240	..Using shift register
241	..Detectors
242	.Current steering
243	..Diode
244	MISCELLANEOUS

D25-1

CLASS D25 BUILDING UNITS AND CONSTRUCTION ELEMENTS

DECEMBER 1987

Excluded from this class are:
Burial vault; see D99-1+.
Cloth sheet material; see D5.
Concrete wall tie; see D8-384.
Diving board; see D21-236.
Elevator; see D34-33.
Escalator; see D34-28.
Fence or railing hardware, e.g., mounting bracket, clip, etc.; see D8.
Fireplace mantel; see D23-404.
Form for flower arrangement; see D11-143+.
Gate hinge; see D8-323+.
Gate latch; see D8-331+.
Lamp combined with post; see D26-68.
Lens or glass for lighting fixture; see D26-120+.
Linoleum rug; see D6-582+.
Linoleum sheet; see D5.
Mobile home or trailer; see D12-101+.
Mobile stair for boarding aircraft, etc.; D34-30.
Pipe, tubing or hose; see D23-266.
Plant stake; see D8-1.
Portable swimmimg pool; see D21-252.
Roof gutter; see D23-267.
Service station canopy combined with pump; see D15-9.1+.
Step plate for vehicle; see D12-203.
Structure which encloses telephone only; see D6-421 and 553+.
Tent; see D21-253.
Tiles assembled into countertop; see D23-308.
Tiles assembled into floor and wall covering; D56-582+.
Track for sliding door; see D8-377.
Vehicle support; see D34-31.
Wallpaper; see D5.

1 STRUCTURE (1)
2 .Swimming pool
3 .Combined with diverse article, e.g., airport complex, etc.
4 .Clustered identical units
5 .Four or more levels
6 ..Tower
7 .Simulative
8 ..Animate
9 ...Humanoid
10 ..Edible article or container therefor
11 ..Vehicle
12 .Arena or stadium
13 .Geodesic dome (2)
14 .Grain storage
15 .Greenhouse
16 .Single occupant type, i.e., telephone booth, privy, ticket booth, etc. (3)
17 .Having intersecting angular or curved roofs (4)
18 .Having curved or vaulted roof
19 ..Dome
20 ..Sinuous
21 ..Concave, catenary or flexibly suspended
22 .Having angular roof
23 ..Trapezoidal or triangular surface
24 ...Gambrel type
25 ...Mansard type
26 ...Pyramidal
27 ...Zig-zag type, i.e., sawtooth
28 ...Inverted
29 ...A-frame type
30 ...Isolated slanted surface
31 .Circular or oval in plan
32 .Polygonal in plan
33 .Fully enclosed
34 ..Having carport or garage
35 PREFABRICATED UNIT (5)
36 .Underground service vault (6)
37 .Elevator cab interior (7)
38 .Railing unit or fence (8)
39 ..Combined, e.g., fence and gate
40 ..Simulative
41 ..Swimming pool grab rail
42 ..Post or board extending above rail, e.g., picket type
43 ..Post and rail type
44 ...Rail centered on post
45 ..Wire type (9)
46 ...Barbed
47 .Closure (10)
48 ..For passage, e.g., door, etc.
49 ...Flexible or folding
50 ...Gate
51 ....Turnstile
52 ..Window
53 ..Screen or guard (11)
54 ...Window well cover
55 .Finishing unit (12)
56 ..Roof or canopy (13)
57 ...Awning type
58 ..Partition, wall or ceiling (25)
59 ...Exterior, e.g., storefront
60 ..Surround for door window, panel, etc. (14)
61 .Framing unit, e.g., building frame (15)
62 STAIR, LADDER, SCAFFOLD OR SIMILAR SUPPORT (16)
63 .Stepped (17)
64 ..Ladder
65 ...Step stool type
66 .Scaffold
67 .Sawhorse
68 .Element or attachment (18)
69 ..Stair tread, step or ladder rung
100 TRELLIS OR TREILLAGE UNIT
101 .Simulative
102 ARCHITECTURAL STOCK MATERIAL (19)
103 .Transparent or translucent
104 ..Simulative
105 ...Plant life
106 ..Three or more pattern repeats about axis on side (20)
107 ...Hollow type
108 ..Hollow type
109 ..Symmetrical on two axes on one side (21)
110 ..Unidirectional repeat pattern
111 ..Random arrangement of pattern elements within repeat unit (22)
112 .Splash block
113 .Construction block or brick type (23)
114 ..Perforated or openwork (24)
115 ...Symmetrical
116 ....Arcuate detail
117 .....Cicular or oval
118 ....Rectangular or square detail
119 .Extruded shape (25)
120 ..Three or more repeats, or uniform configuration, about axis
121 ..Bilaterally symmetrical
122 ...With hollow core
123 ...Parallel repeating ribs or grooves
124 ..With hollow core
125 ..Parallel repeating ribs or grooves
126 .Post or beam
127 ..With cross arm or cantilevered support
128 ..Three or more repats, or uniform configuration, about axis
129 ...Vertically fluted
130 ..Simulative
131 ..With notch or tang for receiving attachment (e.g., fencing wire, etc.)

D25-2

CLASS D25 BUILDING UNITS AND CONSTRUCTION ELEMENTS

DECEMBER 1987

ARCHITECTURAL STOCK MATERIAL (19)
.Post or beam
132 ..Openwork type or with opening
133 ..Element or attachment
134 ...Cross arm or cantilevered support
135 ...Capital or cap
136 .Elongated molding type (e.g., crown, dentil, etc.) (26)
137 ..Simulative
138 .Panel, tile or applique or floor, wall, roof, etc. (27)
139 ..Roofing, or wall shingle
140 ...Tile type
141 ...Stamped, crimped or bent material having uniform thickness
142 ...Perforated or openwork
143 ...Three or more repeats, or uniform configuration, about axis
144 ...Symmetrical on two axes
145 ..Relief carving or medallion type (28)
146 ...Humanoid simulation
147 ...Three or more repeats, or uniform configuration, about axis (e.g., medallion, etc.)
148 ...Symmetrical on two axes
149 ..Simulative, or natural material pattern
150 ...Wood grain
151 ...Brick or stone (29)
152 ..Perforated or openwork (30)
153 ...Symmetrica pattern or pattern unit
154 ....Arcuate
155 .....Circular or oval
156 ....Rectangular or square
157 ..Symmetrical pattern or pattern unit (31)
158 ...Three or more repeats, or uniform configuration, about axis
159 ....Consisting entirely of material having uniform thickness
160 ...On two axes
161 ....Consisting entirely of aterial having uniform thickness
162 ...Consisting entirely of material having uniform thickness
163 ..Textured (e.g., pitted, cratered, drilled, etc)
164 .Elongated (32)
199 MISCELLANEOUS (33)

SEARCH NOTES FOR CLASS D25.

(1) Includes bridge.
For roof with no structure other than support, see subclasses 56+.
(2) For arcuate dome, see subclass 19.
(3) Includes structures which enclose a person.
(4) Search subclasses 56+ for roof, per se.
(5) Includes only plural components which are preassembled as a unit, e.g., steeple, cupola, etc.
For individual components, see subclasses 70+.
For trellis, stairs, louver unit, see, respectively, subclasses 71, 62+ and 87+.
For girder, beam, etc., see subclass 77.
(6) Includes element.
(7) For elevator, see D34-33.
(8) For handrail or post, see respectively, subclasses 73+ and 77.
(9) Includes element.
(10) Includes shutter; includes a combination with surround.
For surround per se, see subclass 60.
(11) Includes only screens or guards which are designed solely for window or door.
For grille adaptable to other architectural uses, see subclasses 87+.
For perforated sheet material, see D92-26+.
(12) Includes rigid drapery cornice.
(13) If structure has wall, see subclasses 1+.
(14) Includes arch structure.
(15) Includes cantilever support.
(16) Includes ramp; includes bleachers.
(17) For mobile stairs, see D34-30.
(18) For component which is adaptable to other article, such as handrail, post, etc., see subclasses 70+.
(19) Classified by full line disclosure
(20) Planar surface not considered.
(21) Planar surface not considered.
(22) If repeat protocol is apparent from the disclosure, patent is classified under earlier subclasses. If, however, the disclosured pattern element does not indicate how it can repeat, patent is classified here.
(23) For edging, see subclass 164.
(24) Opening must extend through the article.
(25) Includes stage set.
(26) If molding has extruded shape, see subclasses 119+.
For corner or end piece, see subclass 102.
(27) For generic elongated stock material having indefinite length, see subclass 164.
(28) For embossed panelwork or for grill, see subclsses 138, 152 or 157 for brick, per se, see subclasses 113+.
(29) Opening must extend through the
(30) Opening must extend through the article. Otherwise patent is classified in subclasses which follow.
(31) Exclusive of random acoustic texture within symmetrical pattern unit, or of tongue or groove edge detail.
(32) Includes railing, edging, etc.
However, if sailing, etc. has an extruded shape, patent is classified in subclasses 119+.
For corner or end piece, see subclass 102.
(33) Includes construction form.
(34) To be classified in this subclass, article must have uninterrupted extruded shape. If openings, protrusions, rivets, vertical striations, etc., are disclosed with an elongated article, patent is classified in subclasses 126, 136 or 164.
For articles with specific end configuration such as fence picket, see subclasses 126+.

PLT-1

CLASS PLT PLANTS

DECEMBER 1986

1 ROSES
2 .Climbers
3 ..White or yellow
4 ..Orange or salmon
5 ..Pink
6 ..Red
7 .Miniatures
8 ..White or yellow
9 ..Pink
10 ..Red
11 .Hybrid teas or grandifloras
12 ..Red bicolor
13 ..Striped colors
14 ..White
15 ..Yellow
16 ..Orange
17 ..Salmon
18 ..Light to medium pink
19 ..Dark pink
20 ..Light to medium red
21 ..Dark red
22 .Floribundas or polyanthas
23 ..White
24 ..Yellow
25 ..Salmon
26 ..Light to medium pink
27 ..Dark pink
28 ..Light to medium red
29 ..Dark red
30 NUTS (INCLUDING ORNAMENTAL VARIETIES)
31 .Pecans
32 .Walnuts
33 FRUITS (EDIBLE AND FLESHY, BUT ALSO INCLUDING ORNAMENTAL VARIETIES)
34 .Apples (including crabapples)
35 ..Sports of "Delicious"
36 .Pears
37 .Cherries
38 .Plums or prunes
39 .Apricots
40 .Nectarines
41 ..Yellow-fleshed
42 .Peaches
43 ..Yellow-fleshed
44 .Avocados
45 .Citrus
46 .Brambles (i.e., the genus Rubus)
47 .Grapes
48 .Strawberries
49 ..Everbearing
50 CONIFERS
51 BROADLEAF TREES
52 .Honey locusts
53 .Poplars
54 SHRUBS OR VINES
55 .Azaleas or rhododendrons
56 ..Light to medium pink
57 ..Dark pink to red
58 .Barberries
59 .Buddleias
60 .Camellias
61 ..Light to medium pink
62 ..Dark pink to red
63 .Euonymus
64 .Flowering quinces
65 .Hollies
66 .Lilacs
67 .English ivies (i. e., Hedera helix varieties)
68 HERBACEOUS FLOWERING PLANTS
69 .African violets
70 .Carnations or pinks
71 ..Light to medium pink
72 ..Dark pink
73 ..Red
74 .Chrysanthemums
75 ..Cushion
76 ..Decorative (i.e., double-flowered)
77 ...White
78 ...Yellow
79 ...Orange
80 ...Light to medium pink
81 ...Dark pink
82 ...Red
83 .Freesias
84 .Fuchsias
85 .Gladioli
86 .Poinsettias
87 .Verbenas
88 HERBACEOUS ORNAMENTAL FOLIAGE PLANTS (INCLUDING LAWN GRASSES)
89 MISCELLANEOUS (E.G., MINTS, HOPS, MUSHROOMS, SUGAR CANE OR TOBACCO)

Appendix 27
Sample Classification Change Order

U. S. DEPARTMENT OF COMMERCE
Patent and Trademark Office

CLASSIFICATION ORDER 1166

FEBRUARY 2, 1988

Project No. Y 8527

The following classification changes will take effect immediately:

	Class	Subclasses	Art Unit	Ex'r Search Room No.
Abolished	250	385	256	CP4 - 8C34
Established	250	385.1; 385.2	256	CP4 - 8C34

No other classes are effected by this project:

Included in this Order are the following documents:

A. CLASSIFICATION MANUAL CHANGES;

B. LISTING OF PRINCIPAL SOURCE OF ESTABLISHED AND DISPOSITION OF ABOLISHED SUBCLASSES;

C. CHANGES TO THE U. S. - I. P. C. CONCORDANCE;

D. DEFINITION CHANGES.

CLASSIFICATION ORDER 1166

February 2, 1988
Project No. Y 8527

A. CLASSIFICATION MANUAL CHANGES

Additional and Modified Subclasses

CLASS 250 RADIANT ENERGY

Classifiers: E. Folsom; L. Bouchard

Examiners : C. Fields; C. Hannaher

Technician : H. Holmes

(EXISTING SUBCLASS) 384 ...Radioactive
385.1 ..Plural chambers or three or more electrodes
385.2 ...Spark chambers

CLASSIFICATION ORDER 1166

FEBRUARY 2, 1988

Project No. Y 8527

B. LISTING OF PRINCIPAL SOURCE OF ESTABLISHED AND DISPOSITION OF ABOLISHED SUBCLASSES

ABOLISHED		DISPOSITION		ESTABLISHED		SOURCE	
Class	Subclass	Class	Subclass	Class	Subclass	Class	Subclass
250	385	250	385.1; 385.2	250	385.1; 385.2	250	385

CLASSIFICATION ORDER 1166

FEBRUARY 2, 1988

Project No. Y 8527

C. CHANGES TO THE U. S. - I. P. C. CONCORDANCE

U. S.		I. P. C.	
Class	Subclass	Subclass	Notation
250	385	DELETE	
250	385.1	G01T	1/18
		H01J	47/00
	385.2		47/10; 47/14

CLASSIFICATION DEFINITIONS

ADDENDA NO. 76 - ORDER NO. 1166

FEBRUARY 2, 1988

D. CHANGES TO THE DEFINITIONS (Project No. Y 8527)

CLASS 250 - RADIANT ENERGY

Definitions Abolished

Subclass

385

Definitions Established

385.1 Plural Chambers or Three or More Electrodes:
Subject matter under subclass 374 wherein there are more than two electrically independent, electrically accessible points or wherein the gas molecules between the electrically accessible points are prevented from free movement between the points by a wall or other barrier.

385.2 Spark Chambers:
Subject matter under subclass 385.1 wherein incoming radiant energy ionizes the gas molecules to trigger a spark between the electrically independent, electrically accessible points.

(1) Note. The position of the spark corresponds to the position of the ionizing event.

SEARCH THIS CLASS, SUBCLASS:
385.1, for multiwire, position-sensitive detectors.

390.12, for position-sensitive neutron detectors.

Appendix 28
Patent for Velcro

United States Patent Office

2,717,437
Patented Sept. 13, 1955

1

2,717,437

VELVET TYPE FABRIC AND METHOD OF PRODUCING SAME

George de Mestral, Prangins, Vaud, Switzerland, assignor to Velcro S. A., Fribourg, Switzerlaud, a corporation of Switzerland

Application October 15, 1952, Serial No. 314,933

Claims priority, application Switzerland October 22, 1951

4 Claims. (C1. 28–72)

My invention has for its object a velvet fabric including a foundation structure constituted by a weft and a warp incorporating threads that are cut at a predetermined length so as to form a raised pile. My novel fabric distinguishes from the other similar fabrics by the fact that the raised pile is made of artificial material, while at least part of the threads in said pile is provided near its end with material-engaging means, as required for adhering to a similar fabric or for scouring purposes.

My invention has for its further object a method for producing a fabric of the above type, according to which the raised pile is provided with its material-engaging means by forming loops round a carrier and submitting the loops formed on the carrier to a thermic action with a view to giving them their final shape, after which the loops are cut on one side of the carrier so that each loop produces at least one pile thread having a hook-shaped end.

2

Fabrics of the type referred to are intended primarily for use as closing means or fasteners for garments, curtains and the like as substitutes for the usual slider-operated closing means or fasteners or for buttons or the like attaching means, whenever a yielding invisible closing arrangement is of advantage.

Fabrics of the type referred to may also be used to advantage as cleaning implements. As a matter of fact, it is possible to lay them on a support made of wood or of plastic material so as to produce a clothes or shoe brush.

I have illustrated diagrammatically and by way of example in accompanying drawings various embodiments of the fabric according to my invention. In said drawings:

Fig. 1 is an explanatory diagram of a preferred method of production of such a fabric.

Fig. 2 shows two pieces of fabric executed according to a first embodiment of my invention and laid over each other so as to interengage and to adhere to each other.

Turning to Fig. 1, it is apparent that the velvet fabric, illustrated in the making, includes a foundation structure constituted by a weft **1** and by a warp **2**.

The foundation structure also carries the warp thread **3** in addition to the warp thread **2**, said thread **3** being adapted to form the raised pile **9, 10**, some of the pile threads showing near their ends material-engaging means; in the example illustrated, the threads **9** of the pile are bent downwardly to form a hook **4**.

Obviously, the weft and warp threads forming the foundation structure may be arranged otherwise than in the manner illustrated.

2,717,437

3

and to make them retain the said desired shape. I may use as an artificial material any suitable plastic material, such as that sold in the trade as "nylon," which is a generic term for any long chain synthetic polymeric amide which has recurring amide groups as an integral part of the main polymer chain and which is capable of being formed into a filament in which the structural elements are oriented in the direction of the axis. Note "Du Pont Products Index," published by E. I. du Pont de Nemours & Company (Inc.), Wilmington 98, Delaware, page 91, January, 1951.

When producing a fabric of the type illustrated in Fig. 1, I proceed in the same manner as for the production of the special valvet made on bar looms. As a matter of fact, it is possible to use for the formation of the pile, small transverse metal bars (Fig. 1) round which the additional warp threads are caused to pass so as to form loops **6**. Each small bar **5** is provided with a longitudinal groove **7** in which is guided a knife **8** adapted to cut the loop **6** open and to form thus the raised pile threads. However, with a view to obtaining the hooks **4**, I heat the bar **5** before the cutting of the loops **6**, so that the thread extending over the bar may assume and retain the shape imparted to it by the latter. The heating of the bars may be obtained by making an electric current flow through them. Obviously, the carrier bars **5** for the loop may be heated as well through any means other than an electric current, e.g. the carrier bars may be hollow and heated by steam.

After the loop **6** has been cut, the raised pile retains its shape and each loop produces, on one hand, the raised pile threads **9**, the ends of which are hook-shaped and, on the other hand, ordinary raised pile threads **10** forming lost strands.

As apparent from inspection of Fig. 2, it is possible to superpose two pieces of fabric of the type illustrated in Fig. 1, after having imparted to one of the two pieces a 90° angular displacement in respect to the other piece and after turning them so that their pile surfaces face each other, the pile threads **9** of

4

one piece engaging the pile threads **9** of the other piece through the co-operating hooks **4**. Thus, as the number of hooks **4** per surface unit, say per square inch, may be high, the two pieces of fabric adhere together perfectly, and it is necessary to draw them away from each other with some energy, when it is desired to separate them. After separation of the two pieces, the hooks **4** return into their original shape.

It is thus possible to use a pair of such pieces of fabric to advantage as a substitute for the usual fastening means, such as slide-operated fasteners, ordinary buttons, press buttons or the like attaching means. As a matter of fact, it is sufficient to sew a piece of fabric of the type described along the edges of the parts of garments, curtains and the like, which are to be held together. A mere pressure exerted on the two garment elements against each other will provide for their fastening. A somewhat considerable tractional stress exerted on the two garment elements thus associated, allows separating them when required.

A fastening arrangement obtained as disclosed hereinabove shows inter alia the following advantages:

The possibility of compensating any clearance between the associated elements as such elements are not always in exact register with each other;

In the case of any straining, the fastening arrangement will yield before any damage is inflicted on the fabric, which is very important whenever a piece of velvet has engaged some fabric having delicate meshes.

It should be remarked that the velvet fabrics according to my invention and more particularly those illustrated in Figs. 1 and 2, may serve advantageously for the execution of household implements, such as clothes brushes, shoe brushes and the like cleaning or scouring means. Obviously, in such a case, the size of the threads and more particularly their thickness and their rigidity may be selected according to the purpose intended for the pieces of fabric that are to be executed.

The velvet fabric according to my invention is thus obtained in practice in a manner similar to a conventional velvet. However, it is obvious that my novel fabric has neither the silky feel nor the outer appearance of the usual velvet that serves for the execution of clothes or for upholstery.

I claim:

1. A method for producing a velvet type fabric consisting in weaving together a plurality of weft threads and a plurality of warp threads together with a plurality of auxiliary warp threads of synthetic resin material, forming loops with said auxiliary warp threads on one surface of the so woven fabric, submitting the said loops to a thermal source, thereby causing said loops to retain their shape to form raised pile threads, cutting said loops near their outer ends, thereby forming material-engaging means on at least a portion of said pile threads constituted by said cut loops.

2. A method for producing a velvet type fabric consisting in weaving together a plurality of weft threads and a plurality of warp threads together with a plurality of auxiliary warp threads of synthetic resin material, forming loops with said auxiliary warp threads on one surface of the so woven fabric, submitting the said loops to a thermal source, thereby causing said loops to retain their shape to form raised pile threads, cutting each of said loops near the respective outer end at a point between said outer end and the fabric surface, thereby forming a hook-shaped section with the free end of the respective pile thread at one side of said point at which the cut is made.

3. A velvet type fabric comprising a foundation structure including a plurality of weft threads, a plurality of warp threads, and a plurality of auxiliary warp threads of a synthetic resin material in the form of raised pile threads, the ends of at least part of said raised pile threads being in the form of material-engaging hooks.

4. A velvet type fabric comprising a foundation structure including a plurality of weft threads, a plurality of warp threads, and a plurality of auxiliary warp threads of a synthetic resin material in the form of raised pile threads, the terminal portions of at least part of said raised pile threads being in the form of a material-engaging means including hook-shaped sections.

5

10

15

20

25

References Cited in the file of this patent

UNITED STATES PATENTS

2,062,884 Holland.........Dec. 1, 1936
2,662,559 Miller..........Dec. 15, 1953

Appendix 29
Differences between United States and Foreign Patent Laws

Major Differences Between United States and Foreign Patent Laws.

	United States	Foreign
Prior Art	One year grace period.	Absolute novelty - Anything publicly known prior to filing date is prior art.
Term	17 years from grant.	20 years from filing.
Publication	Only at grant.	18 months after priority filing date, again when allowable, and again at grant.
Examination	Conducted in all cases.	Deferred until examination fee is paid. 5 - 7 years after filing.
Applicant	Inventor.	Owner or inventor.
Fees	Filing, Issue, and 3 maintenance fees	Filing, Search, Examination, and annual maintenance fees after third year.

U.S. PATENT AND TRADEMARK OFFICE

3

Major Differences Between United States and Foreign Patent Laws (continued)

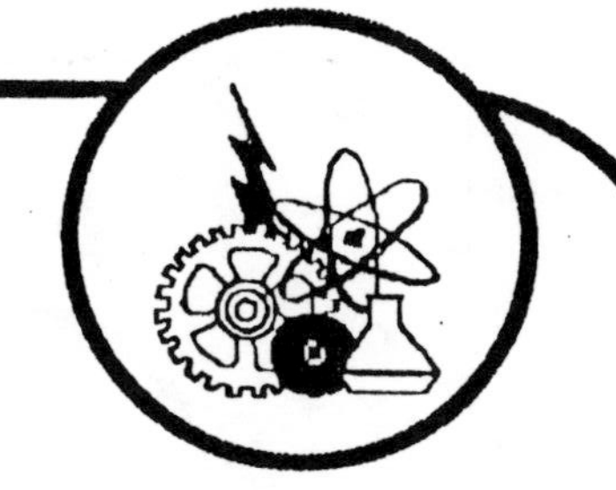

	United States	Foreign
Kinds of Patents	Utility Design Plant Division Continuation Continuation-in-part Reissue Reexamination Certificate Statutory Invention Registration	Utility Petty (utility model) Inventor's Certificate Division Patents of Addition
Working Requirements	None	Some Countries
Mandatory Licensing	None	Some Countries
Conflicting Applications	Interference	Patent issued to First to file

Major Differences Between United States and Foreign Patent Laws (continued)

	United States	Foreign
Opposition	Reexamination limited to prior patents and printed publications	Usually 3 to 4 months after notice of allowance for public to oppose grant on any grounds
Representation	Inventor may represent himself or herself	Non-resident applicant must be represented by a local agent

Major Multi-national Patent Treaties

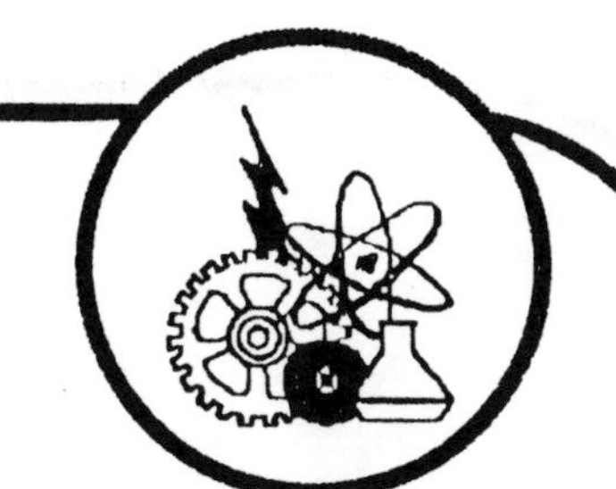

Paris Convention for the Protection of Industrial Property

Originally signed on March 20, 1883. Has been revised 6 times thereafter. 98 member States. Copy in Appendix P of MPEP.

Relates to equal treatment for foreign and domestic nationals, provides for foreign priority benefit and other topics.

Inter-American Convention relating to Inventions, Patents, Designs and Industrial Models

Signed August 20, 1910. Relates to national treatment and foreign priority benefits.

Patent Cooperation Treaty

Signed in Washington June 19, 1970, 43 member States.

Provides for filing a single international patent application with effect in designated States. Provides for an international search, publication, and international preliminary examination.

Convention for the Protection of New Varieties of Plants (UPOV)

Signed 1961, 17 member States.

Relates to national treatment and right of priority in protection of plants.

Major Multi-national Patent Treaties (continued)

Budapest Treaty on the International Recognition of the Deposit of Microorganisms for the Purposes of Patent Procedure.

Agreed to in 1977, 22 member States.

Relates to deposit of microorganisms necessary for full disclosure in patent applications.

European Patent Convention (EPC)

Signed in 1973, Came into effect on June 1, 1978, 14 member States.

Provides for one application, search, examination and grant of a patent with effect in designated European States.

Community Patent Convention (CPC)

Agreed to in 1975. Not yet in force. Provides for a single patent being issued for all European Economic Community (EEC) countries. Requires ratification by all EEC States.

African Intellectual Property Organization (OAPI)

13 Member States. Organization of French speaking African nations.

Headquarters at Yaounde, Cameroon.

Major Multi-national Patent Treaties (continued)

African Regional Industrial Property Organization (ARIPO)
11 member States, Agreed 1976. Headquarters Harara, Zimbabwe.

Convention Establishing the World Intellectual Property Organization
119 member States. Ensures administrative cooperation among the intellectual property unions, States, and international organizations.

Strasbourg Agreement Concerning the International Patent Classification
27 Member States, agreed October 7, 1975. Concerns International Patent Classification System.

European Patent Office

<u>Facilities</u>

- The Hague, Netherlands

  Formalities review and searching

- Berlin

  Some searching of German language applications, printing

- Munich, Federal Republic of Germany

  Examination as to patentability

  Boards of Appeal

  Administration

European Patent Convention Procedures

- Application filed with designations of countries in which patent protection is desired.
- Application may be filed in English, French, or German, or in any official language of a member State.
- Application checked for completeness and informalities.
- Records and publication front page prepared.
- Prior art search conducted.
- Application published 18 months after priority date. (translations required of claims into other two official languages)
- Examination must be requested within six months of the publication.
- Annual fees must be paid beginning at the third year of pendency to the EPO.
- Patent issues with effect in all designated countries (translations of patent required into official languages of most designated States.
- Public has 9 months after publication of patent to protest grant of patent.
- Patent must be enforced by courts in each designated country.
- Annual fees must be paid to each designated country to maintain patent in force.

11 U.S. PATENT AND TRADEMARK OFFICE

Japanese Patent Office

New patent laws came into effect in 1971 and 1987.

Application must be filed in Japanese language.

Application may be for either a patent or a utility model.

All applications are laid open to public at 18 months after priority date.

Provisional protection with damages at level of reasonable royalty for infringement after application has been laid open.

Full disclosure of patent application is printed.

Brief explanation of drawing, drawing and claim published for utility model applications.

Examination may be deferred until 7 years date after filing for patent applications, until 4 years after filing date for utility models.

Request for examination may be made by applicant or a third party or by the patent office

If examination is not requested within 7 or 4 year period, application is withdrawn.

When allowable, application is published for opposition.

Public has three months to file opposition.

Term of patent is 15 years from date of publication for opposition but cannot exceed 20 years from filing date.

Appendix 30
Class 160 from *Manual of Classification*

CLASS 160 CLOSURES, PARTITIONS AND PANELS, FLEXIBLE AND PORTABLE

DECEMBER 1988

1 AUTOMATIC CONTROL
2 .Non-thermal automatic initiator
3 ..Force initiated
4 ...Weight
5 ..Weather and/or light initiated
6 .Non-fusible thermal initiator
7 .With starting or driving means
8 .With retarding means
9 .With releasing means for operator and/or counterbalance
10 WITH SIGNAL, INDICATOR OR SIGN
11 WITH WIPER OR CLEANER
12 WITH INSECT EXITS
13 .Applied to slidably interconnected frames or panels
14 .Zigzag or crimped surface
15 .Spaced or overlapping sections
16 .Conical, prismatic or other protuberances
17 .Bar, grooved or apertured
19 WITH HOOD, CANOPY, SHIELD STORAGE CHAMBER OR OUTRIGGED RIGID PANEL
20 .Outrigged rigid panel, with flexible panel sides
21 .Changeable size
22 .For outrigged flexible panels (i.e., awnings)
23.1 .For roll type
24 ..Portable
25 ..With plural flexible or portable panels
26 ..Casement housed roll
27 ...Fabric free edge connected to movable closure
28 ..Fabric free edge connected to movable closure
29 ..Movable hood, canopy, shield, storage chamber or outrigged rigid panel
30 ...Mounted upon movable closure
31 ..Combined with frame or demountable side guides
32 .For plural strip, slat or panel and/or pleating type
33 ..Casement housed
34 ..Venetian blind and/or collapsing
35 ..Pleating or edge hinged gathering type
36 ..Edge hinged, slidably mounted
37 .Slidable into storage chamber
38 .Structure
39 ..Extensible
40 WITH FILLER AND FILLER TYPE
41 .For roll type
42 ..Between outrigged roll type
43 .Between parallel plane, relatively slidable panels
44 WITH LIQUID SUPPLYING AND/OR DRAINING MEANS
45 WITH OUTRIGGERS (I.E., AWNINGS)
46 .Floor or ground engaging outrigger
47 .With groove engaging mounting means →
48 .With rigid closure
49 ..Movable rigid closure related
50 .With non-outrigged flexible panel
51 ..Convertible
52 ..With non-outrigged roll type
53 .Umbrella type
54 .Plural outrigged type
55 ..One roll type
56 .Non-planar fabric arrangements
57 ..Planar surface, with sides
58.1 ...Sides foldable, rollable, or collapsible
59 .Multi-positional
60 .Convertible
61 .Plural strip, slat or panel type
62 ..Telescopic and/or collapsible
63 .With fabric having diverse areas
64 .Adjustable size
65 .With braced outrigger
66 .Roll type
67 ..Shiftable position roll
68 ..Fabric free edge connected operator
69 ..Multi-part outrigger
70 ...Two part, intermediate pivot
71 ...Telescopic
72 ..Swinging outriggers
73 ...With pivotal motion preventing means
74 ...Sliding pivot
75 ..Sliding outriggers
76 .With fabric frame
77 ..Pivoted
78 .Multi-part outrigger
79 ..Two part, intermediate pivot
80 ..Telescopic
81 .Swinging
82 ..With sliding pivot
83.1 .Rigid or nonmovable
84.1 PLEATING TYPE
84.2 .Lazy-tong links pivot about axes transverse to panel
84.3 .Lazy-tong links pivot about axes parallel to panel
85 PLURAL RUN TYPE
86 .Endless
87 PLURAL AND/OR WITH RIGID CLOSURE
88 .Non-planar, non-parallel arrangement
89 .Diverse types
90 ..With rigid closure
91 ...Diverse rigid closures
92 ...Swinging rigid closures
93 ....Intermediate axis
94 .....With opposite, parallel, offset flexible panels
95 ....Flexible or portable panel related to rigid closure operating or fastening means
96 ....Facing flexible or portable panel and mounted for optional or conjoint movement
97 ....Flexible or portable panel complementary to form complete closure
98 ...Roll type flexible panel
99 ....Connected to slidable rigid closure
100 .....Fabric free edge connected closure
101 ...Flexible panel between slidable parallel plane rigid closures
102 ...Flexible panel connected to and moved by slidable rigid closure
103 ...Flexible or portable panel related to rigid closure operating or fastening means
104 ...Grille or shutter type
105 ...Flexible panel removably mounted in slidable rigid closure guides
106 ...Removably mounted over fabric in frame
107 ...Plural strip, slat or panel type
108 ..Roll and hanging or drape type only
109 ...Shiftable position
110 ....With extensible pole, roller, bar or support therefor
111 ....Vertically slidable
112 ...With extensible pole, roller, bar or support therefor
113 .Plural strip, slat or panel assemblies
114 ..Non-planar
115 ..Multiple section unit
116 ...One closing passage through another
117 ..Mounted on opposite sides of single opening
118 ...Track guided
119 ..Mounted on opposite sides of single support
120 .Roll type
121.1 ..Single roll

CLASS 160 CLOSURES, PARTITIONS AND PANELS, FLEXIBLE AND PORTABLE

DECEMBER 1988

	PLURAL AND/OR WITH RIGID CLOSURE
	.Roll type
122	..Differently directed fabrics
123	.Hanging or draped type
124	..Side-by-side arranged, on single pole or or track
125	..Shiftable pole
126	..Laterally overlapping fabrics and/or co-elevational parallel tracks
127	COMBINED
128	CONVERTIBLE
129	.Plural movably interconnected strip, slat or panel type
130	PLURAL STRIP, SLAT OR PANEL TYPE
131	.Longitudinally crooked strip, slat or panel
132	..Bow or U-shape, pivoted together at ends
133	.Roll type
134	.Fan type
135	.Portable
136	.Parallelogram type
137	..Plural differently movable parallelogram section with a common side
138	..With operator
139	...Unit collapsing only
140	....Diverse types of operators
141	.....One operator a spring
142	.....One operator a weight
143	......Weighted parallelogram element extension
144	....Spring
145	.....In brace
146	....Weight
147	.....Strand carried
148	.....Weighted parallelogram element extension
149	....With gearing
150	....With strand and pulley
151	....Unit carried operator
152	..With means to hold unit in partially collapsed or collapsed position
153	...Combined tension and compression means
154	...Pivoted compression bar
155	....Friction catch
156	...Tension members
157	....Flexible member foreshortened at end
158	....Friction catch
159	..Unit collapsing on guides
160	..Unit pivoted to support
161	..Unit structures
162	...With additional set of plural strips, slats or panels
163	....Plural diverse additional sets
164	....Additional set composed of interconnected strips, slats or panels
165	....Pivots parallel to unit plane
166.1	.Venetian blind type
167	..Changeable position assemblies
168.1	..With accumulating means
169	...Accumlated at stationary end
170	...Drum or roller wound strand
171	....Axially traversed drum or roller
172	...With side guides
173	...Units and subcombinations thereof
174	..Adjustable strip, slat or panel angle
175	...With latch or detent
176.1	...With angle adjusting means
177	...Units and subcombinations thereof
178.1	..Subcombination or structure thereof
178.2	...Cord lock
178.3	...Slat support (e.g., tape ladder)
179	.Plural fabrics in a single frame
180	.One closing opening in another
181	.With mounting or supporting means
182	..Removable frame type support
183	..Non-planar non-parallel arrangement on support
184	..Strip, slat or panel not interconnected for relative motion
185	..Strip, slat or panel interconnected for diverse relative motions
186	..Arranged to provide plural passageways on opening
187	...Strip, slat or panel pivotally interconnected
188	..With operating means
189	...Counterbalance and additional operator
190	...Counterbalance only
191	....Spring
192	...Spring
193	...Strand
194	..Movably mounted track
195	...Pivotally mounted
196.1	..Unit hung from horizontal track
197	...Strips, slats or panels slidably interconnected
198	...Track parallel to axis of interconnection pivots
199	...Strip, slat or panel constrained for pivotal folding
200	...With independently movable strip, slat or panel
201	..Track guided
202	...Strips, slats or panels slidably interconnected
203	...Unit pivoted to support, pivot displaced from unit plane
204	...Track parallel to interconnection pivot axis
205	...With independently movable strip, slat or panel
206	...Strip, slat or panel constrained for pivotal folding
207	....Vertical track
208	...A single strip, slat or panel connected to track
209	...With means to force against mount or support
210	..Pivotally mounted unit
211	...Strips, slats or panels slidably interconnected
212	...Support pivot at angle to interconnection pivot axis
213	...Constrained for pivotal folding
214	..Slidably mounted unit
215	..Removable unit
216	...Strip, slat or panel slidably interconnected
217	...Pivotal interconnection necessary to removal and insertion
218	.Movably interconnected
219	..Unit having diverse areas
220	..Interconnected for plural relative motions
221	...Doubly extensible by sliding only
222	..Slidably interconnected
223	...Plural individually interconnected strips, slats or panels
224	...With operator and/or anti-friction means
225	...With relative slidable motion preventing means
226	...Embracing type
227	....Integral embracing means
228	...Tongue, bead, bolt or dog into groove or slot
229.1	..Edge-to-edge interconnected
230	...Single covering fabric
231.1	...Interconnected by strand, fabric, rubber, or plastic
231.2	....Interconnected by rubber or plastic

CLASS 160 CLOSURES, PARTITIONS AND PANELS, FLEXIBLE AND PORTABLE

DECEMBER 1988

	PLURAL STRIP, SLAT OR PANEL TYPE
	.Movably interconnected
	..Edge-to-edge interconnected
232	...Hollow, filled or covered elements
233	...With relative pivotal motion preventing means
234	....Acting to keep unit planar only
235	...By curled or bent side edges
236	.Strip or slat structure
237	WITH FABRIC HAVING DIVERSE AREAS
238	ROLL TYPE
239	.Framed
240	..Adjustable size frame
241	.Plural roll
242	.Shiftable position roll
243	..Roll translated for winding or unwinding operation only
244	...Fabric free edge adjustable
245	...Spring roller
246	..Straight line non-axial translation in fabric plane only
247	...Center operated, roll support
248	...Rack and pinion operated
249	...One side, non-strand operated
250	...Extensible roller and/or roll support
251	...Step-by-step
252	...Strand shifted
253	....Pulley or drum lift on roller
254	....Winding drum
255	.....With roll guide
256	....Roll intermediate guides for translating strands
257	....With position holding means
258	.....Strand holder
259	......Roll guided
260	....Roll guided
261	...With elevator rod
262	.Crooked roll, non-cylindrical or flexible roller
263	.Adjustable dimension roll
264	.With fabric reinforcements
265	.With fabric free edge connected operator
266	.With guides or fabric edge holders
267.1	..Fabric side edge and stick
268.1	..Fabric side edge
269	...Clamping
270	...Slot or channel type
271	....Fabric receiving
272	.....Guide mounted in a channel
273.1	.....With interlock between fabric and guide
274	..Stick guided only
275	...With positive stop or detent
276	....Reciprocable detent
277	...Guides of bar or strand form
278	....With release mechanism
279	....Guide strand axially through or into stick
280	...Leading edge holders movable axially of stick
281	....With release mechanism
282	.....With guide engaging rollers and friction elements
283	......And roll brake
284	.....With guide engaging rollers
285	....With guide engaging rollers and friction elements
286	.....And roll brake
287	....With guide engaging rollers
288	....With axially acting friction elements
289	...With friction elements
290.1	.With fabric leading edge fastening means
291	.With brake or stop
292	..Plural
293.1	..Limit of travel
294	...For spring operated roller
295	....Screw operated brake or stop
296	..Speed limiting
297	..Ball or roller, form brake or stop
298	..Friction
299	...For spring operated roller
300	..Pawl or detent
301	...For spring operated roller
302	....Spring urged pawl or detent
303	....Pawl or detent mounted off the roller
304.1	....Sliding pawl or detent
305	..For spring operated roller
306	...With means to prevent uncoiling of spring on removal from support
307	..For strand operated roll
308	...Strand operated brake or stop
309	.With rotating means
310	..Electric operation or control
311	..Motor or fluid pressure control
312	..Plural, diverse
313	..Spring
314	...External of roll
315	...With winding or tensioning means
316	...Nested helical springs
317	...Plural
318	...With spring anchor
319	..Strand
320	...With weight or handgrip
321	...Endless
322	...With compensating and shock absorbing spring
323.1	.With supporting or journaling means or with roller end structure
324	..With pintle clamping or holding means
325	..With pintle journaled in rolled
326	..With roller gudgeons
327	NON-FRAMED PLURAL EDGED HELD FABRIC
328	.With fabric stretching means
329	..Resilient stretching means
330	HANGING OR DRAPE TYPE
331	.Motor operating means or electric or fluid pressure control
332	.Chain, cable and/or strand type
333	.Shiftable position pole
334	..Plural motions
335	..Swingable
336	...Vertical swing
337	...With pole operating means
338	..Vertically shiftable
339	...With strand operator
340	.With fabric operating means
341	..Laterally gathered fabric
342	...Lazy tongs
343	...Spring or screw
344	...Strand
345	....Hollow or slotted track
346	.....With anti-friction means
347	....With anti-friction means
348	.With pleating means
349.1	.With bottom or intermediate holding, weighting, or draping means
349.2	..Intermediate holding means
350	.Portable
351	PORTABLE
352	NON-PLANAR (E.G., THREE DIMENSION)
353	WITH PLURAL FRAMES
354	WITH NON-RIGID FRAME OR BORDER ELEMENTS
368.1	WITH MOUNTING, FASTENING, OR SUPPORTING MEANS
368.2	.Shiftable support for overlying shield, e.g., pillow sham type
369	.For framed type
370	.With guide for strand
370.1	.Grain car type temporary closure
370.2	.Automobile windshield weather protector or glare shield
371	FRAMED TYPE

CLASS 160 CLOSURES, PARTITIONS AND PANELS, FLEXIBLE AND PORTABLE

DECEMBER 1988

	FRAMED TYPE
372	.Adjustable frame size
373	..With excess or extensible fabric
374	..Double extensible
374.1	...Expandable at corner joint, e.g., artist's canvas stretcher frame
375	..Shiftable edge bar section only
376	...With spring thrust
377	.Collapsible or knockdown
378	.Fabric stretching
379	.With intermediate reinforcing bars, members or braces
380	.Two superimposed frame elements, fabric clamping
381	.With corner structure
382	FABRIC FASTENING MEANS
383	.To elongated element
384	..Additional things fastened
385	..With modified fabric
386	...Plural layers in panel portion
387	...With fabric hem
388	...With loops
389	...With strips or strands
390	...With fabric openings or pockets
391	..Longitudinally divided or with fabric receiving externally exposed channel or groove
392	...Internal fastener larger than groove or channel fabric exit
393	....With end cap and/or sleeve
394	....Sheet material elongated element
395	...Elongated fastener in channel or groove
396	....With end cap and/or sleeve
397	....Sheet material elongated element
398	...With pointed, piercing and/or hook elements
399	...Externally applied clamp, clasp, sleeve or end cap
400	..Sheet or strand
401	..By sleeve and/or end cap
402	..By externally applied clamp or clasp
403	..By elongated fastener
404	..By pointed, piercing and/or hook elements
405	MISCELLANEOUS AND PROCESSES OF USING

****

CROSS-REFERENCE ART COLLECTIONS

****

900	VERTICAL TYPE VENETIAN BLIND
901	LAZY-TONG CONNECTED PLURAL STRIPS, SLATS, OR PANELS
902	VENETIAN BLIND TYPE BRACKET MEANS
903	ROLL TYPE BRACKET MEANS
904	ELECTRIC OR PNEUMATIC AWNING OPERATOR
905	LAZY-TONG-LINK AWNING OPERATOR
906	SCREW-THREADED AWNING OPERATOR
907	SPRING (OTHER THAN SPRING ROLLER) AWNING OPERATOR
908	STRAND AWNING OPERATOR
909	.ENDLESS STRAND
910	.DRUM WOUND STRAND
911	WORM GEAR AWNING OPERATOR
912	RACK AND PINION AWNING OPERATOR
913	GEAR AWNING OPERATOR

****

DIGESTS

****

DIG 1	Auto radiator screens
DIG 2	Auto screens and miscellaneous
DIG 3	Auto visors and glare shields
DIG 4	Auto side awnings
DIG 5	Awning head and front bars
DIG 6	Bath curtains
DIG 7	Fabric
DIG 8	Flexible door
DIG 9	Fireplace screen
DIG 10	Roll screen
DIG 11	Roll screen idler roll
DIG 12	Overlapping, on windows
DIG 13	Suction cup
DIG 14	Step joints
DIG 15	Web-to-tube fasteners
DIG 16	Magnetic
DIG 17	Venetian blinds, motor driven
DIG 18	Zipper
DIG 19	Storm sash

Appendix 31
Description for Class 160

Class 160, CLOSURES, PARTITIONS AND PANELS, FLEXIBLE AND PORTABLE

CLASS DEFINITIONS

I. This is the generic class for:

(a) Devices in the form of one or more flexible panel units (see section II);

(b) Devices in the form of panel units made up for plural strips, slats, or panels interconnected for relative motion (see section III);

(c) Devices in the form of panel units or partitions, including those completely rigid and/or portable, having means combined therewith for facilitating the passage of insects therethrough in one direction and preventing or making their passage difficult in the reverse direction (see subclass 12 and the notes thereunder);

(d) Devices in the form of portable partitioning panel units (see subclass 351 and the notes thereunder);

(e) Combinations of (a) through (d) with each other, with rigid panel units, including rigid closures, or with other structure not elsewhere provided for (see subclass 127 and the notes thereunder).

II. Flexible panel units are considered to be all those in which a flexible fabric or other flexible sheet material forms the panel portion, even though it may have a rigid frame, rigid enforcements, rigid support means for one or more edges thereof, or combinations of these features.

III. Panel units made up of plural strips, slats or panels interconnected for relative motion include all units in which means are provided for directly connecting adjacent elements together to provide for relative motion therebetween. The individual strip, slat or panel elements may be rigid. The relative motion provided may be, for example, one or more relative swinging or pivotal motions, relative sliding motions, or combinations of such motions.

Panel units made up of plural rigid strips, slats or panels are not considered to be interconnected for relative motion and are, therefore, excluded:

(a) where a single strip, slat or panel is movably connected to a second strip, slat or panel which is immovably mounted or supported;

(b) where the only connection therebetween is a common operator and/or a common support.

Such common support may be a single frame having either or both (1) rigid strips, slats or panels immovably mounted thereon, (2) rigid strips, slats or panels movably mounted thereon and connected only through the frame and/or by common operating means.

IV. These devices are termed, for example, closure, partition, weather protective, ventilation, ornamentation and like devices, more specifically called awnings, curtains, shades, blinds, gates, doors, windows, shutters, screens, roll tops for desks, picture screens, fire place screens, photographic backgrounds, vehicle storm fronts, shields, visors, robes or aprons, etc.

V. Since this is the generic class, it takes the subjects matter specified in sections I to III only when of a character not provided for in some other main class, for which see the notes below.

VI. The devices classified in this class and detailed in sections I-III, supra, have been generically termed flexible and portable panels. In contradistinction, those closures and panels not of the flexible panel type, have generically been called rigid closures and rigid panels. (For the classification of closures in general, see Class 220, Metallic Receptacles, subclasses 200+ and the notes thereto).

NOTES

The following notes are those considered necessary to stand out in connection with the main class definition. Many of them are repeated in connection with subclass definitions, and other notes to other main classes occur only under the particular subclasses to which they are pertinent.

(1) Note. *Combinations*. In general, art devices of which a flexible or portable panel is but a part are classified with the appropriate art and cross-referenced in this class when the disclosure of structure of the panel, its mounting and/or operating means warrants the same.

However, combinations of flexible or portable panels as above defined, with other closures or panels (herein called rigid closures or rigid panels) are in this class, subclasses 19+ and subclasses 87+.

(2) Note. *Enclosures*. This class does not have flexible or portable panels related to each other or to other structure to form enclosures. For such devices, see the class appropriate to the type of enclosure claimed, as the various receptacle, cabinet, building structure, tent, vehicle, etc., classes; and also Class 128, Surgery, subclass 372. Enclosures should be distinguished from this class (160) subclasses 19+ and other nonplanar flexible and portable panel arrangements such as are in subclasses 45+ (including subclass 53 and subclasses 56+), subclasses 86, 88, 131+, 183, 262 and 352.

(3) Note. *Support Subcombinations*. Support subcombinations for flexible and portable panels of the hanging and drape type are classified as follows:

See Classes:

16, Miscellaneous Hardware, subclasses 87+. These subclasses take: (A) nonmovably mounted tracks, both per se and in combination with their supporting means; (B) travelers per se (except as set forth in notes (4) to (6), subclass 87.2) and in combination with the track as in (A); and (C) either (A) or (B) in combination with the hanging fabric or other flexible sheet material where only those features thereof are claimed as pertain to connection with the travelers. Features claimed in addition to those enumerated above (such as movable mounting of track, fabric operating means, additional characteristics of the fabric, etc.) cause classification in Class 160, Closures, Partitions and Panels, Flexible and Portable.

211, Supports, Racks, particularly subclasses 87+. These subclasses take the structure of both single and plural rods or other elongated elements claimed in combination with mounting means (including those of the bracket type) constructed to movably support the elongated element and having operating means, and also takes nonmovably mounted single or plural rods where features or rod structure in addition to the connection between the rod and bracket are claimed (not pertaining to the means for fastening the hanging or drape fabric). Features in addition to the above, claiming the fabric either broadly or specifically, means to fasten the fabric to the elongated element, means to operate the fabric, etc. cause classification in Class 160, Closures, Partitions and Panels, Flexible and Portable or Class 16, Miscellaneous Hardware.

. . .

79. Devices under subclass 78 in which the outrigger consists of two parts pivoted together at an intermediate point.

SEARCH THIS CLASS, SUBCLASS:
70, for this device with roll type structure, whether or not claimed.

80. Devices under subclass 78 in which the outrigger consists of two or more telescopic members which may be extended longitudinally in a straight line.

SEARCH THIS CLASS, SUBCLASS:
71, for this device with disclosed roll type structure whether or not claimed.

81. Devices under subclass 45 in which the outrigger is pivoted to the support.

82. Devices under subclass 81 wherein the pivot has a sliding motion along the support.

SEARCH THIS CLASS, SUBCLASS:
74, for this device with disclosed roll type structure, whether or not claimed.

83.1 Rigid or Nonmovable:
Device under subclass 45 in which the outrigger is rigid or nonmovable.

SEARCH THIS CLASS, SUBCLASS:
46, for such an outrigger which engages the ground.

84.1 PLEATING TYPE:
Device under the class definition in which the fabric may be contracted by imparting folds thereto or may be expanded by straightening or tending to straighten such folds, where means are provided to cause such folds to occur at or along predetermined lines by engaging the fabric at points spaced across the fabric or by elongated elements at the fold lines.

(1) Note. For devices of the hanging or drape type which are freely suspended from their upper edge, and (a) which have folds imparted thereto by lateral gathering or other means, see subclasses 341+ and 348; or (b) which have other gathering means, see subclass 340.

SEARCH THIS CLASS, SUBCLASS:
130+, (and see the notes thereunder particularly subclass 230 for related structures in which strips or slats are associated together.

SEARCH CLASS:
92, Expansible Chamber Devices, subclasses 34+ for a bellows type expansible chamber device.
105, Railway Rolling stock, subclasses 18+ for a bellows type diaphragm between two articulated railroad cars.

84.2 Lazy-Tong Links Pivot About Axes Transverse to Panel:
Pleating type device under subclass 84.1 wherein the means causing or straightening the folds of the fabric, during the contraction or extension thereof, consists of a lazy-long type extensible framework in which a series of jointed bars engaging the fabric pivot about axes perpendicular to the plane formed by the fabric whenever it is fully extended.

85. Devices under the class definition in which a single flexible panel is looped over or under at least one pole, or other guide roller to cover the same or portions of the same area, and has means for moving the flexible panel relative to such pole roller or other guide.

. . .

95. Devices under subclass 92 in which the flexible or portable panel has some feature designed to coact with or otherwise be related to the operating or fastening means of the rigid closure, e.g., apertures through which the handle passes.

SEARCH THIS CLASS, SUBCLASS:
103, for corresponding devices associated with other types of rigid closure.

96. Devices under subclass 92 in which the flexible or portable panel and the rigid closure or rigid panel are mounted in facing relation and are so mounted that they can be moved either separately or together.

97. Devices under subclass 92 in which the flexible or portable panel closes a part of an opening and the rigid closure or rigid panel closes the remainder of the opening.

98. Devices under subclass 90 in which the rigid closure or rigid panel is combined with a roll type flexible panel.

99. Combinations under subclass 98 in which some interconnection between a roll type and a rigid sliding closure is claimed.

(1) Note. For the most part the roll is (1) mounted on a sliding closure or (2) so interconnected therewith as to follow its motion.

SEARCH THIS CLASS, SUBCLASS:
29+, for such devices having a hood, canopy, shield or storage chamber.
242+ particularly 243+ and 246+ for other movably mounted devices of the roll type.

100. Combinations under subclass 99 in which the free or leading edge of material as it unrolls is connected to the rigid sliding closure.

SEARCH THIS CLASS, SUBCLASS:
27 and 28 for free edge-fastening means, when fastened to a movable closure.
290.1+, for such devices when not so fastened.

101. Combinations under subclass 90 in which the flexible panel is mounted in any way between two or more sliding rigid closures.

102. Devices under subclass 90 in which the flexible panel is connected to and caused to move by slidable movable closure. The connection may be through operating means that transmits motion between the two.

103. Devices under subclass 90 in which the flexible or portable panel has some feature designed to coact with or otherwise be related to the operating or fastening means of the rigid closure, e.g., apertures through which the handle passes.

SEARCH THIS CLASS, SUBCLASS:
95, for other flexible or portable panels.

104. Devices under subclass 90 in which the rigid closure or panel is of the grille or shutter type.

105. Devices under subclass 90 in which the flexible or rigid panel is removably mounted in or on the guides that guide the rigid-closure or panel.

SEARCH THIS CLASS, SUBCLASS:
47, for outrigged devices disclosed for this combination.
375+, for frame type devices with shiftable side bars, many of which are disclosed for this combination.

106. Devices under subclass 90 in which the flexible panel is in the form of a fabric in a frame and the rigid closure or panel is mounted (usually removably) on or in the frame and over the fabric.

107. Devices under subclass 90 in which the flexible panel is of the plural strip, slat or panel type in which (1) a plurality of strips, slats and/or panels utilizing flexible material in their construction are claimed in combination or (2) a plurality of rigid strips, slats and/or panels are interconnected with each other for relative motion and form a single unit, or (3) both.

108. Diverse devices of two types only under subclass 89, one being of the roll type and one of the hanging or drape type.

. . .

SEARCH CLASS:

108, Horizontally Supported Planar Surfaces, subclasses 11+ for a horizontally supported planar surface convertible to a flexible panel.

182, Fire Escapes, Ladders, Scaffolds, subclasses 21+ for a ladder convertible to a shutter.

211, Supports, Racks, subclasses 2+ for support racks convertible to flexible and portable panels.

312, Supports, Cabinet Structures, subclasses 3+ for cabinets convertible to flexible and portable panels.

129. Devices under subclass 128 in which the device is of the plural strip, slat or panel type, and the strips, slats or panels are interconnected for relative motion, as set forth in paragraph III of the main class definition.

130. Devices under the class definition in which (1) a plurality of strips, slats and/or panels utilizing flexible material in their construction are claimed in combination or (2) a plurality of rigid strips, slats and/or panels are interconnected with each other for relative motion and form a single unit, or (3) both.

(1) Note. Plural rigid strips, slats or panels interconnected only by a common operator or support (including a frame) will be found in Class 49, Movable or Removable Closures and Class 98, Ventilation.

SEARCH THIS CLASS, SUBCLASS:

39, for similar subject matter disclosed as a hood, canopy, shield or storage chamber for a flexible or portable panel device.

84.1+ and 85+, for related subject matter as defined therein.

113+, for the combination of a plurality of units as defined in this subclass (130); and see the notes to said subclass (113) for the stated line between these two subclasses.

123+, (and the notes to subclass 123), for a plurality of separate hanging or drape type flexible panels.

332, for a plurality of chain cables or strands which hang from a support.

349, for bottom holding means, *per se*, even if disclosed with a device for this subclass (130).

353, for plural frames for single flexible panels.

371+, for framed or panel type devices; usually distinguished from this subclass (130) in that only a single fabric is used.

372+, for adjustable frames for single flexible panels, including those slidably interconnected U-shaped frame members with a single flexible fabric.

375+, for single frame type devices with slidable rigid edge sections, such devices not having been treated as plural strip, slat or panel type.

379, for those single frame type devices with intermediate reinforcing bars and members where there is but a single fabric for the frame.

380, for single fabrics clamped between two frame elements.

SEARCH CLASS:

49, Movable or Removable Closures, subclasses 73+ for closures interconnected for concurrent movement. See (1) Note above.

52, Static Structures, e.g., Buildings, subclasses 660+ for rigid fabric or lattice openwork, e.g., slatted floor covering.

296, Land Vehicles, Bodies and Tops, subclasses 210+ for land vehicles having top with openings therein and movable panels as closures therefor.

428, Stock Material or Miscellaneous Articles, subclasses 44+, 53, 54+ and 57+ for plural, joined webs or sheets which make up stock material, rather than a "unit" of the type referred to in the definition of subclass 130.

131. Devices under subclass 130 wherein at least one strip, slat or panel has its major axis in other than a straight line.

SEARCH THIS CLASS, SUBCLASS:
32+, where one strip, slat or panel houses another.
57+ and subclasses 61+ for outrigged flexible panels of the plural strip, slat and panel type with separate side units not a part of the main panel.

132. Devices under subclass 131 in which plural strips, slats or panels, each of bow or U-shape, have the ends of corresponding arms adjacent and pivoted together.

133. Plural strip, slat or panel devices under subclass 130 in which the strips, slats or panels are accumulated in the form of a roll.

(1) Note. Many of these devices have no roller upon which they are accumulated, the end bar, strip, slat or panel being the element about which the others are rolled.

SEARCH THIS CLASS, SUBCLASS:
121.1+, for single roll having panels wound thereupon in superimposed or overlapping relation so that at least portions of adjacent panels wind up together.
238+, for other roll type devices and see the notes thereunder.

134. Devices under subclass 130 wherein at least two of the elements are arranged either in the same plane or in parallel planes so that they may swing around a single actual or virtual pivot which is at right angles to such plane or planes.

135. Devices under subclass 130 having means peculiarity adapted to make the device portable including those having legs or bases to support the device in upright position.

SEARCH THIS CLASS, SUBCLASS:
351, and see the note thereto for other portable devices.

136. Devices under subclass 130 wherein two intersecting sets of parallel strips, slats or panels are interconnected by pivots at their intersections.

(1) Note. The parallelogram type of device may be a mere operator for another type of plural strip, slat or panel device.

137. Devices under subclass 136 wherein at least two sections of a parallelogram unit are pivotally interconnected through a common side, and which sections move differently when the device is operated. The different movements may take place concurrently and usually consist of pivotal movements in opposite directions.

138. Devices under subclass 136 having an operator for either moving the device with respect to its mounting means and/or collapsing the device.

(1) Note. Counterbalances have been considered as operators within this definition.

(2) Note. Where both moving and collapsing operators are claimed, the patent is placed here and cross referenced into subclasses 139+.

139. Devices under subclass 138 wherein the operator effects the collapsing of the device only and does not move the unit as a whole with respect to its support.

(1) Note. Since the majority of parallelogram type devices expand in one direction when they collapse in another, the term "collapse" has been used generically in referring to any motion about the parallelogram pivots.

140. Devices under subclass 139 where there are two diverse types of operators.

141. Devices under subclass 140 wherein one of the operators is a spring.

SEARCH THIS CLASS, SUBCLASS:
144+, for other spring type operators.

142. Devices under subclass 140 wherein one of the operators is a weight.

SEARCH THIS CLASS, SUBCLASS:
146+, for other weight type operators.

143. Devices under subclass 142 wherein an extension of one of the parallelogram elements is weighted.

SEARCH THIS CLASS, SUBCLASS:
148, for additional weighted parallelogram extensions.

144. Devices under subclass 139 wherein the operator is a spring.

145. Devices under subclass 144 wherein the spring is part of a brace for the parallelogram unit.

146. Devices under subclass 139 wherein the operator is a weight.

147. Devices under subclass 146 wherein the weight is carried by a strand.

SEARCH THIS CLASS, SUBCLASS:
150, for other operators utilizing a strand and pulley.

148. Devices under subclass 146 wherein an extension of one of the parallelogram elements is weighted.

SEARCH THIS CLASS, SUBCLASS:
143, for this feature with disclosed diverse type of operator whether or not claimed.

149. Devices under subclass 139 wherein the operator includes gearing to either change the mechanical advantage of the operator, or to allow the operator to be controlled from a remote point.

150. Devices under subclass 139 wherein the operator includes a strand and pulley.

151. Devices under subclass 139 wherein the operator is completely carried by the unit.

152. Devices under subclass 136 having means to hold the unit in either partially or completely collapsed position.

153. Devices under subclass 152 wherein the holding means comprises both tension and compression means.

154. Devices under subclass 152 wherein the holding means is a pivoted compression bar.

155. Devices under subclass 154 wherein the compression bar is frictionally fastened in adjustable position.

SEARCH THIS CLASS, SUBCLASS:
158, for similar catches for tension members.

156. Devices under subclass 152 wherein the holding means is a tension member.

157. Devices under subclass 156 wherein a flexible end of a tension member is foreshortened to change the effective length of the member.

158. Devices under subclass 156 wherein the tension member is frictionally fastened in adjustable position.

SEARCH THIS CLASS, SUBCLASS:
155, for similar friction devices on compression members.

159. Devices under subclass 136 wherein the unit collapses on guides.

160. Devices under subclass 136 wherein the unit is pivoted to its support.

161. Parallelogram structures as defined in subclass 136 wherein no structure of the operator or mounting means is claimed.

162. Devices under subclass 161 wherein at least one additional set of plural strips, slats or panels is mounted on the device for movement therewith.

163. Devices under subclass 162 wherein there are a plurality of diverse sets of plural strips, slats or panels.

164. Devices under subclass 162 wherein one additional set is composed of interconnected strips, slats or panels.

165. Devices under subclass 162 wherein the pivots of the parallelogram type are parallel to the plane of the unit as a whole.

Appendix 32
Sample CASSIS CD-ROM Search for Sunscreens

CASSIS CD-ROM screen showing input of patent number for retrieval of classes and title.

F1:Help F2:Browse F3:Display F4:Connection F5:Storage F6:Setup F7:Quit

SEARCH

U.S. Department of Commerce
Patent and Trademark Office

PATENT BIBLIOGRAPHIC FILE - 1969 TO DATE

1

Patent Number: 4202396
Issue Year:
Assignee Code:
State or Country:
Status:
Classification:
Title or Abstract:

Use arrow keys to highlight search field.

Touch ENTER to start search, CTRL-BREAK to abort search.

Connection: Total: 1	

The marked number was found.

F1:Help F2:Full/List F3:Format (Sort) F5:Output F6: Jump F7:Done

	Patent Title Listing: 48 of 64
Patent Number	Patent Title
4466474	ADJUSTING DEVICE FOR A SLAT BLIND CONTAINED IN A SE
4459778	VENETIAN BLIND HANGER PIVOT ASSEMBLY
4458740	DEVICE FOR ROTATING A SUNSHADE STRIP ENCLOSED IN A
• 4202396	MOTOR VEHICLES AND SUNSHIELDS

Screen showing bibliographic information. Note that because of the data, no abstract was available on CD-ROM. Abstract could be found in the OG or by using a commercial data base.

F1:Help F2:Full/List F3:Format (Sort) F5:Output F6:Jump F7:Done

	Patent Display: 1 of 1
Patent Number	4202396
Issue Year	980
Assignee Code	0
State / Country	ILX
Classification	160/107 160/84.1 160/DIG2 160/229.1 296.97.7 296/97.8
Title	MOTOR VEHICLES AND SUNSHIELDS

Screen from CASSIS CD-ROM where the selected class/subclass combination was entered.

F1:Help F2:Browse F3:Display F4:Connection F5: Storage F6:Setup F7:Quit

SEARCH

U.S. Department of Commerce
Patent and Trademark Office

PATENT BIBLIOGRAPHIC FILE - 1969 TO DATE

Patent Number:
Issue Year:
Assignee Code:
State or Country:
Status:
Classification: 160/107
Title or Abstract:

Use arrow keys to highlight search field.

Touch ENTER to start search, CTRL-BREAK to abort search.

Connection: Total:

Search on class 160 sub 107 retrieved 64 entries; these were looked at in reverse order until...

F1:Help	F2:Full List	F3:Format (Sort)	F5:Output	F6: Jump	F7:Done

Patent Number	Patent Title — Patent Title Listing: 64 of 64
3703920	VENETIAN BLIND AND INSTALLATION
3552473	VENETIAN BLIND
3454073	WINDOW UNIT WITH POLL-TYPE SLATS
3443624	VENETIAN BLIND WINDOW

Appendix 33
List of U.S. Patent Numbers and Their Time Frames

U.S. Patent Numbers and Their Time Frames

YEAR		YEAR	
1988	4,716,594 -	1965	3,163,865 - 3,226,728
1987	4,633,526 - 4,716,593	1964	3,116,487 - 3,163,864
1986	4,562,596 - 4,633,525	1963	3,070,801 - 3,116,486
1985	4,490,885 - 4,562,595	1962	3,013,103 - 3,070,800
1984	4,423,523 - 4,490,884	1961	2,966,681 - 3,013,102
1983	4,366,579 - 4,423,522	1960	2,919,443 - 2,966,680
1982	4,308,622 - 4,366,578	1959	2,866,973 - 2,919,442
1981	4,242,757 - 4,308,621	1958	2,818,567 - 2,866,972
1980	4,180,867 - 4,242,756	1957	2,775,762 - 2,818,566
1979	4,131,952 - 4,180,866	1956	2,728,913 - 2,775,761
1978	4,065,812 - 4,131,951	1955	2,698,434 - 2,728,912
1977	4,000,520 - 4,065,811	1954	2,664,562 - 2,698,433
1976	3,930,271 - 4,000,519	1953	2,624,046 - 2,664,561
1975	3,858,241 - 3,930,270	1952	2,580,379 - 2,624,045
1974	3,781,914 - 3,858,240	1951	2,536,016 - 2,580,378
1973	3,707,729 - 3,781,913	1950	2,492,944 - 2,536,015
1972	3,631,539 - 3,707,728	1949	2,457,797 - 2,492,943
1971	3,551,909 - 3,631,538	1948	2,433,824 - 2,457,796
1970	3,487,470 - 3,551,908	1947	2,413,675 - 2,433,823
1969	3,419,907 - 3,487,469	1946	2,391,856 - 2,413,674
1968	3,360,800 - 3,419,906	1945	2,366,154 - 2,391,855
1967	3,295,143 - 3,360,799	1944	2,338,081 - 2,366,153
1966	3,226,729 - 3,295,142	1943	2,307,007 - 2,338,080

YEAR	
1942	2,268,540 - 2,307,006
1941	2,227,418 - 2,268,539
1940	2,185,170 - 2,227,417
1939	2,142,080 - 2,185,169
1938	2,104,004 - 2,142,079
1937	2,066,309 - 2,104,003
1936	2,026,516 - 2,066,308
1935	1,985,878 - 2,026,515
1934	1,941,449 - 1,985,877
1933	1,892,663 - 1,941,448
1932	1,839,190 - 1,892,662
1931	1,787,424 - 1,839,189
1930	1,742,181 - 1,787,423
1929	1,696,897 - 1,742,180
1928	1,654,521 - 1,696,896
1927	1,612,700 - 1,654,520
1926	1,568,040 - 1,612,699
1925	1,521,590 - 1,568,039
1924	1,478,996 - 1,521,589
1923	1,440,362 - 1,478,995
1922	1,401,948 - 1,440,361
1921	1,364,063 - 1,401,947
1920	1,326,899 - 1,364,062
1919	1,290,027 - 1,326,898
1918	1,251,458 - 1,290,026
1917	1,210,389 - 1,251,457
1916	1,166,419 - 1,210,388

YEAR	
1915	1,123,212 - 1,166,418
1914	1,083,267 - 1,123,211
1913	1,049,326 - 1,083,266
1912	1,013,095 - 1,049,325
1911	980,178 - 1,013,094
1910	945,010 - 980,177
1909	908,436 - 945,009
1908	875,679 - 908,435
1907	839,799 - 875,678
1906	808,618 - 839,798
1905	778,834 - 808,617
1904	748,567 - 778,833
1903	717,521 - 748,566
1902	690,385 - 717,520
1901	664,827 - 690,384
1900	640,167 - 664,826
1899	616,871 - 640,166
1898	596,467 - 616,870
1897	574,369 - 596,466
1896	552,502 - 574,368
1895	531,619 - 552,501
1894	511,744 - 531,618
1893	488,976 - 511,743
1892	466,315 - 488,975
1891	443,987 - 466,314
1890	418,665 - 443,986
1889	395,305 - 418,664

YEAR	
1888	375,720 - 395,304
1887	355,291 - 375,719
1886	333,494 - 355,290
1885	310,163 - 333,493
1884	291,016 - 310,162
1883	269,820 - 291,015
1882	251,685 - 269,819
1881	236,137 - 251,684
1880	223,211 - 236,136
1879	211,078 - 223,210
1878	198,733 - 211,077
1877	185,813 - 198,732
1876	171,641 - 185,812
1875	158,350 - 171,640
1874	140,120 - 158,349
1873	134,504 - 140,119
1872	122,304 - 134,503
1871	110,617 - 122,303
1870	98,460 - 110,616
1869	85,503 - 98,459
1868	72,959 - 85,502
1867	60,658 - 72,958
1866	51,784 - 60,657
1865	45,685 - 51,783
1864	41,047 - 45,684
1863	37,266 - 41,046
1862	34,045 - 37,265

YEAR	
1861	31,005 - 34,044
1860	28,842 - 31,004
1859	22,477 - 28,841
1858	19,010 - 22,476
1857	16,324 - 19,009
1856	14,009 - 16,323
1855	12,117 - 14,008
1854	10,358 - 12,116
1853	9,512 - 10,357
1852	8,622 - 9,511
1851	7,865 - 8,621
1850	6,981 - 7,864
1849	5,993 - 6,980
1848	5,409 - 5,992
1847	4,914 - 5,408
1846	4,348 - 4,913
1845	3,873 - 4,347
1844	3,395 - 3,872
1843	2,901 - 3,394
1842	2,413 - 2,900
1841	1,923 - 2,412
1840	1,465 - 1,922
1839	1,061 - 1,464
1838	546 - 1,060
1837	110 - 545
1836	1 - 109

Appendix 34
Comparing Utility Patents, Design Patents, Trademarks, and Copyrights

Comparing Utility Patents, Design Patents, Trademarks, and Copyrights

	Subject Matter	Time Limit	Infringement	Agency
Utility Patent	machines, articles of manufacture, compositions of matter,	17 years	making, using or selling	US PTO
Design Patent	ornamental designs for articles of manufacture	14 years	making, using or selling	US PTO
Plant Patent	asexually reproduced plants	14 years	making, using or selling	US PTO
Trademark	words, names, symbols or combinations used to distinguish	20 years renewable	likelihood of confusion or deception	US PTO
Copyright	writings, music, works of art that are the original expressions of an idea	life of the author + 50 yrs	literal copying	Office of Copyright Library of Congress

Index